Pitman Research Notes in Mathematics Series

Submission of proposals for consideration
Suggestions for publication, in the form of outlines and representative samples, are invited by the Editorial Board for assessment. Intending authors should approach one of the main editors or another member of the Editorial Board, citing the relevant AMS subject classifications. Alternatively, outlines may be sent directly to the publisher's offices. Refereeing is by members of the board and other mathematical authorities in the topic concerned, throughout the world.

Preparation of accepted manuscripts
On acceptance of a proposal, the publisher will supply full instructions for the preparation of manuscripts in a form suitable for direct photo-lithographic reproduction. Specially printed grid sheets are provided and a contribution is offered by the publisher towards the cost of typing. Word processor output, subject to the publisher's approval, is also acceptable.

Illustrations should be prepared by the authors, ready for direct reproduction without further improvement. The use of hand-drawn symbols should be avoided wherever possible, in order to maintain maximum clarity of the text.

The publisher will be pleased to give any guidance necessary during the preparation of a typescript, and will be happy to answer any queries.

Important note
In order to avoid later retyping, intending authors are strongly urged not to begin final preparation of a typescript before receiving the publisher's guidelines and special paper. In this way it is hoped to preserve the uniform appearance of the series.

Longman Scientific & Technical
Longman House
Burnt Mill
Harlow, Essex, UK
(tel (0279) 26721)

Titles in this series

1 Improperly posed boundary value problems
A Carasso and A P Stone

2 Lie algebras generated by finite dimensional ideals
I N Stewart

3 Bifurcation problems in nonlinear elasticity
R W Dickey

4 Partial differential equations in the complex domain
D L Colton

5 Quasilinear hyperbolic systems and waves
A Jeffrey

6 Solution of boundary value problems by the method of integral operators
D L Colton

7 Taylor expansions and catastrophes
T Poston and I N Stewart

8 Function theoretic methods in differential equations
R P Gilbert and R J Weinacht

9 Differential topology with a view to applications
D R J Chillingworth

10 Characteristic classes of foliations
H V Pittie

11 Stochastic integration and generalized martingales
A U Kussmaul

12 Zeta-functions: An introduction to algebraic geometry
A D Thomas

13 Explicit *a priori* inequalities with applications to boundary value problems
V G Sigillito

14 Nonlinear diffusion
W E Fitzgibbon III and H F Walker

15 Unsolved problems concerning lattice points
J Hammer

16 Edge-colourings of graphs
S Fiorini and R J Wilson

17 Nonlinear analysis and mechanics: Heriot-Watt Symposium Volume I
R J Knops

18 Actions of fine abelian groups
C Kosniowski

19 Closed graph theorems and webbed spaces
M De Wilde

20 Singular perturbation techniques applied to integro-differential equations
H Grabmüller

21 Retarded functional differential equations: A global point of view
S E A Mohammed

22 Multiparameter spectral theory in Hilbert space
B D Sleeman

24 Mathematical modelling techniques
R Aris

25 Singular points of smooth mappings
C G Gibson

26 Nonlinear evolution equations solvable by the spectral transform
F Calogero

27 Nonlinear analysis and mechanics: Heriot-Watt Symposium Volume II
R J Knops

28 Constructive functional analysis
D S Bridges

29 Elongational flows: Aspects of the behaviour of model elasticoviscous fluids
C J S Petrie

30 Nonlinear analysis and mechanics: Heriot-Watt Symposium Volume III
R J Knops

31 Fractional calculus and integral transforms of generalized functions
A C McBride

32 Complex manifold techniques in theoretical physics
D E Lerner and P D Sommers

33 Hilbert's third problem: scissors congruence
C-H Sah

34 Graph theory and combinatorics
R J Wilson

35 The Tricomi equation with applications to the theory of plane transonic flow
A R Manwell

36 Abstract differential equations
S D Zaidman

37 Advances in twistor theory
L P Hughston and R S Ward

38 Operator theory and functional analysis
I Erdelyi

39 Nonlinear analysis and mechanics: Heriot-Watt Symposium Volume IV
R J Knops

40 Singular systems of differential equations
S L Campbell

41 N-dimensional crystallography
R L E Schwarzenberger

42 Nonlinear partial differential equations in physical problems
D Graffi

43 Shifts and periodicity for right invertible operators
D Przeworska-Rolewicz

44 Rings with chain conditions
A W Chatters and C R Hajarnavis

45 Moduli, deformations and classifications of compact complex manifolds
D Sundararaman

46 Nonlinear problems of analysis in geometry and mechanics
M Atteia, D Bancel and I Gumowski

47 Algorithmic methods in optimal control
W A Gruver and E Sachs

48 Abstract Cauchy problems and functional differential equations
F Kappel and W Schappacher

49 Sequence spaces
W H Ruckle

50 Recent contributions to nonlinear partial differential equations
H Berestycki and H Brezis

51 Subnormal operators
J B Conway

52 Wave propagation in viscoelastic media
F Mainardi
53 Nonlinear partial differential equations and their applications: Collège de France Seminar. Volume I
H Brezis and J L Lions
54 Geometry of Coxeter groups
H Hiller
55 Cusps of Gauss mappings
T Banchoff, T Gaffney and C McCrory
56 An approach to algebraic K-theory
A J Berrick
57 Convex analysis and optimization
J-P Aubin and R B Vintner
58 Convex analysis with applications in the differentiation of convex functions
J R Giles
59 Weak and variational methods for moving boundary problems
C M Elliott and J R Ockendon
60 Nonlinear partial differential equations and their applications: Collège de France Seminar. Volume II
H Brezis and J L Lions
61 Singular systems of differential equations II
S L Campbell
62 Rates of convergence in the central limit theorem
Peter Hall
63 Solution of differential equations by means of one-parameter groups
J M Hill
64 Hankel operators on Hilbert space
S C Power
65 Schrödinger-type operators with continuous spectra
M S P Eastham and H Kalf
66 Recent applications of generalized inverses
S L Campbell
67 Riesz and Fredholm theory in Banach algebra
B A Barnes, G J Murphy, M R F Smyth and T T West
68 Evolution equations and their applications
F Kappel and W Schappacher
69 Generalized solutions of Hamilton-Jacobi equations
P L Lions
70 Nonlinear partial differential equations and their applications: Collège de France Seminar. Volume III
H Brezis and J L Lions
71 Spectral theory and wave operators for the Schrödinger equation
A M Berthier
72 Approximation of Hilbert space operators I
D A Herrero
73 Vector valued Nevanlinna Theory
H J W Ziegler
74 Instability, nonexistence and weighted energy methods in fluid dynamics and related theories
B Straughan
75 Local bifurcation and symmetry
A Vanderbauwhede
76 Clifford analysis
F Brackx, R Delanghe and F Sommen
77 Nonlinear equivalence, reduction of PDEs to ODEs and fast convergent numerical methods
E E Rosinger
78 Free boundary problems, theory and applications. Volume I
A Fasano and M Primicerio
79 Free boundary problems, theory and applications. Volume II
A Fasano and M Primicerio
80 Symplectic geometry
A Crumeyrolle and J Grifone
81 An algorithmic analysis of a communication model with retransmission of flawed messages
D M Lucantoni
82 Geometric games and their applications
W H Ruckle
83 Additive groups of rings
S Feigelstock
84 Nonlinear partial differential equations and their applications: Collège de France Seminar. Volume IV
H Brezis and J L Lions
85 Multiplicative functionals on topological algebras
T Husain
86 Hamilton-Jacobi equations in Hilbert spaces
V Barbu and G Da Prato
87 Harmonic maps with symmetry, harmonic morphisms and deformations of metrics
P Baird
88 Similarity solutions of nonlinear partial differential equations
L Dresner
89 Contributions to nonlinear partial differential equations
C Bardos, A Damlamian, J I Díaz and J Hernández
90 Banach and Hilbert spaces of vector-valued functions
J Burbea and P Masani
91 Control and observation of neutral systems
D Salamon
92 Banach bundles, Banach modules and automorphisms of C*-algebras
M J Dupré and R M Gillette
93 Nonlinear partial differential equations and their applications: Collège de France Seminar. Volume V
H Brezis and J L Lions
94 Computer algebra in applied mathematics: an introduction to MACSYMA
R H Rand
95 Advances in nonlinear waves. Volume I
L Debnath
96 FC-groups
M J Tomkinson
97 Topics in relaxation and ellipsoidal methods
M Akgül
98 Analogue of the group algebra for topological semigroups
H Dzinotyiweyi
99 Stochastic functional differential equations
S E A Mohammed

100 Optimal control of variational inequalities
V Barbu
101 Partial differential equations and dynamical systems
W E Fitzgibbon III
102 Approximation of Hilbert space operators. Volume II
C Apostol, L A Fialkow, D A Herrero and D Voiculescu
103 Nondiscrete induction and iterative processes
V Ptak and F-A Potra
104 Analytic functions – growth aspects
O P Juneja and G P Kapoor
105 Theory of Tikhonov regularization for Fredholm equations of the first kind
C W Groetsch
106 Nonlinear partial differential equations and free boundaries. Volume I
J I Díaz
107 Tight and taut immersions of manifolds
T E Cecil and P J Ryan
108 A layering method for viscous, incompressible L_p flows occupying R^n
A Douglis and E B Fabes
109 Nonlinear partial differential equations and their applications: Collège de France Seminar. Volume VI
H Brezis and J L Lions
110 Finite generalized quadrangles
S E Payne and J A Thas
111 Advances in nonlinear waves. Volume II
L Debnath
112 Topics in several complex variables
E Ramírez de Arellano and D Sundararaman
113 Differential equations, flow invariance and applications
N H Pavel
114 Geometrical combinatorics
F C Holroyd and R J Wilson
115 Generators of strongly continuous semigroups
J A van Casteren
116 Growth of algebras and Gelfand–Kirillov dimension
G R Krause and T H Lenagan
117 Theory of bases and cones
P K Kamthan and M Gupta
118 Linear groups and permutations
A R Camina and E A Whelan
119 General Wiener–Hopf factorization methods
F-O Speck
120 Free boundary problems: applications and theory, Volume III
A Bossavit, A Damlamian and M Fremond
121 Free boundary problems: applications and theory, Volume IV
A Bossavit, A Damlamian and M Fremond
122 Nonlinear partial differential equations and their applications: Collège de France Seminar. Volume VII
H Brezis and J L Lions
123 Geometric methods in operator algebras
H Araki and E G Effros
124 Infinite dimensional analysis–stochastic processes
S Albeverio
125 Ennio de Giorgi Colloquium
P Krée
126 Almost-periodic functions in abstract spaces
S Zaidman
127 Nonlinear variational problems
A Marino, L Modica, S Spagnolo and M Degiovanni
128 Second-order systems of partial differential equations in the plane
L K Hua, W Lin and C-Q Wu
129 Asymptotics of high-order ordinary differential equations
R B Paris and A D Wood
130 Stochastic differential equations
R Wu
131 Differential geometry
L A Cordero
132 Nonlinear differential equations
J K Hale and P Martinez-Amores
133 Approximation theory and applications
S P Singh
134 Near-rings and their links with groups
J D P Meldrum
135 Estimating eigenvalues with *a posteriori/a priori* inequalities
J R Kuttler and V G Sigillito
136 Regular semigroups as extensions
F J Pastijn and M Petrich
137 Representations of rank one Lie groups
D H Collingwood
138 Fractional calculus
G F Roach and A C McBride
139 Hamilton's principle in continuum mechanics
A Bedford
140 Numerical analysis
D F Griffiths and G A Watson
141 Semigroups, theory and applications. Volume I
H Brezis, M G Crandall and F Kappel
142 Distribution theorems of L-functions
D Joyner
143 Recent developments in structured continua
D De Kee and P Kaloni
144 Functional analysis and two-point differential operators
J Locker
145 Numerical methods for partial differential equations
S I Hariharan and T H Moulden
146 Completely bounded maps and dilations
V I Paulsen
147 Harmonic analysis on the Heisenberg nilpotent Lie group
W Schempp
148 Contributions to modern calculus of variations
L Cesari
149 Nonlinear parabolic equations: qualitative properties of solutions
L Boccardo and A Tesei
150 From local times to global geometry, control and physics
K D Elworthy

151 A stochastic maximum principle for optimal control of diffusions
U G Haussmann
152 Semigroups, theory and applications. Volume II
H Brezis, M G Crandall and F Kappel
153 A general theory of integration in function spaces
P Muldowney
154 Oakland Conference on partial differential equations and applied mathematics
L R Bragg and J W Dettman
155 Contributions to nonlinear partial differential equations. Volume II
J I Díaz and P L Lions
156 Semigroups of linear operators: an introduction
A C McBride
157 Ordinary and partial differential equations
B D Sleeman and R J Jarvis
158 Hyperbolic equations
F Colombini and M K V Murthy
159 Linear topologies on a ring: an overview
J S Golan
160 Dynamical systems and bifurcation theory
M I Camacho, M J Pacifico and F Takens
161 Branched coverings and algebraic functions
M Namba
162 Perturbation bounds for matrix eigenvalues
R Bhatia
163 Defect minimization in operator equations: theory and applications
R Reemtsen
164 Multidimensional Brownian excursions and potential theory
K Burdzy
165 Viscosity solutions and optimal control
R J Elliott
166 Nonlinear partial differential equations and their applications. Collège de France Seminar. Volume VIII
H Brezis and J L Lions
167 Theory and applications of inverse problems
H Haario
168 Energy stability and convection
G P Galdi and B Straughan
169 Additive groups of rings. Volume II
S Feigelstock
170 Numerical analysis 1987
D F Griffiths and G A Watson
171 Surveys of some recent results in operator theory. Volume I
J B Conway and B B Morrel
172 Amenable Banach algebras
J-P Pier
173 Pseudo-orbits of contact forms
A Bahri
174 Poisson algebras and Poisson manifolds
K H Bhaskara and K Viswanath
175 Maximum principles and eigenvalue problems in partial differential equations
P W Schaefer
176 Mathematical analysis of nonlinear, dynamic processes
K U Grusa
177 Cordes' two-parameter spectral representation theory
D F McGhee and R H Picard
178 Equivariant K-theory for proper actions
N C Phillips
179 Elliptic operators, topology and asymptotic methods
J Roe
180 Nonlinear evolution equations
J K Engelbrecht, V E Fridman and E N Pelinovski
181 Nonlinear partial differential equations and their applications. Collège de France Seminar. Volume IX
H Brezis and J L Lions
182 Critical points at infinity in some variational problems
A Bahri
183 Recent developments in hyperbolic equations
L Cattabriga, F Colombini, M K V Murthy and S Spagnolo
184 Optimization and identification of systems governed by evolution equations on Banach space
N U Ahmed
185 Free boundary problems: theory and applications. Volume I
K H Hoffmann and J Sprekels
186 Free boundary problems: theory and applications. Volume II
K H Hoffmann and J Sprekels
187 An introduction to intersection homology theory
F Kirwan
188 Derivatives, nuclei and dimensions on the frame of torsion theories
J S Golan and H Simmons
189 Theory of reproducing kernels and its applications
S Saitoh
190 Volterra integrodifferential equations in Banach spaces and applications
G Da Prato and M Iannelli
191 Nest algebras
K R Davidson
192 Surveys of some recent results in operator theory. Volume II
J B Conway and B B Morrel
193 Nonlinear variational problems. Volume II
A Marino and M K Murthy
194 Stochastic processes with multidimensional parameter
M E Dozzi
195 Prestressed bodies
D Iesan
196 Hilbert space approach to some classical transforms
R H Picard
197 Stochastic calculus in application
J R Norris
198 Radical theory
B J Gardner
199 The C* – algebras of a class of solvable Lie groups
X Wang

200 Stochastic analysis, path integration and dynamics
D Elworthy
201 Riemannian geometry and holonomy groups
S Salamon
202 Strong asymptotics for extremal errors and polynomials associated with Erdös type weights
D S Lubinsky
203 Optimal control of diffusion processes
V S Borkar
204 Rings, modules and radicals
B J Gardner
205 Numerical studies for nonlinear Schrödinger equations
B M Herbst and J A C Weideman
206 Distributions and analytic functions
R D Carmichael and D Mitrović
207 Semicontinuity, relaxation and integral representation in the calculus of variations
G Buttazzo
208 Recent advances in nonlinear elliptic and parabolic problems
P Bénilan, M Chipot, L Evans and M Pierre
209 Model completions, ring representations and the topology of the Pierce sheaf
A Carson
210 Retarded dynamical systems
G Stepan
211 Function spaces, differential operators and nonlinear analysis
L Paivarinta
212 Analytic function theory of one complex variable
C C Yang, Y Komatu and K Niino
213 Elements of stability of visco-elastic fluids
J Dunwoody
214 Jordan decompositions of generalised vector measures
K D Schmidt
215 A mathematical analysis of bending of plates with transverse shear deformation
C Constanda
216 Ordinary and partial differential equations Volume II
B D Sleeman and R J Jarvis
217 Hilbert modules over function algebras
R G Douglas and V I Paulsen
218 Graph colourings
R Wilson and R Nelson
219 Hardy-type inequalities
A Kufner and B Opic
220 Nonlinear partial differential equations and their applications. College de France Seminar. Volume X
H Brezis and J L Lions

Hilbert modules over function algebras

Ronald G Douglas
SUNY at Stony Brook
and
Vern I Paulsen
University of Houston

Hilbert modules over function algebras

Longman Scientific & Technical
Copublished in the United States with
John Wiley & Sons, Inc., New York

Longman Scientific & Technical,
Longman Group UK Limited,
Longman House, Burnt Mill, Harlow
Essex CM20 2JE, England
and Associated Companies throughout the world.

Copublished in the United States with
John Wiley & Sons, Inc., 605 Third Avenue, New York, NY 10158

First published 1989

AMS Subject Classification: (Main) 47D25, 47A45, 46H25
(Subsidiary) 46J10, 46M20, 471320

ISSN 0269-3674

British Library Cataloguing in Publication Data
Douglas, Ronald G.
Hilbert modules over function algebras.
1. Operator algebras
I. Title II. Paulsen, Vern I., 1951–
512'.55

ISBN 0-582-04796-X

Library of Congress Cataloging-in-Publication Data
Douglas, Ronald G.
Hilbert modules over function algebras/Ronald G. Douglas and Vern I. Paulsen.
p. cm.– (Pitman research notes in mathematics series; 217)
Includes index.
ISBN 0-470-21478-3
1. Function algebras. 2. Hilbert modules. I. Paulsen, Vern I., 1951– .
II. Title. III. Series.
QA326.D69 1989
512'.55—dc20 89-8218
CIP

Printed and bound in Great Britain
by Biddles Ltd, Guildford and King's Lynn

Contents

To Bunny and Sue -
our best collaborators.

Preface

Since first learning the model theory of Sz-Nagy and Foiaş for contractions in the middle sixties, I have sought a more conceptual framework for the theory. In particular, I sought a context in which the transition from one to several commuting operators would be possible. Early attempts yielded first a more (Hilbert Space) geometrical approach and then a development in which the role of the disk algebra was emphasized. Although this allowed the extension of model theory to other planar algebras, even a beginning for the case of several variables proved elusive. In the early seventies, however, several events involving the algebraic notion of a module came together for me to suggest that this might be the key.

First, there was the book of Adams and Griffiths which reported on Serre's "GAGA paper" showing that two very different aspects of complex spaces, one analytic and the other algebraic, were the same when expressed in the language of modules. Second, J. Taylor introduced an elegant notion of joint spectrum in operator theory using modules and obtained very nice results. Finally, Johnson and others showed that Hochschild cohomology, a theory also based on modules, was extremely useful in the study of operator algebras. Based on all of this as well as on conversations with Taylor, Johnson and many others, I became convinced that a module approach to model theory in operator theory would be worthwhile. However, except for an isolated collaboration with Foiaş in the middle seventies, I allowed this idea to lay dormant for over a decade. Some of the motivation for my research during this period, however, was derived from this program, especially my joint work with M. Cowen which introduced complex geometric methods into operator theory.

About four years ago, I began actively developing this theme, partly in collaboration with Paulsen and Misra. My discussions with Paulsen were especially crucial in making the notions of module tensor product and localization central in the development. An invitation by Shunhua Sun to deliver a series of lectures at Szechuan University in September, 1985, gave me the opportunity to provide a systematic overview of the theory. Further progress was made in connection with a series of lectures I delivered in July, 1986, at the University of Victoria for the annual Canadian Conference on Operator Theory/Operator Algebras. These research notes are a substantial revision and extension of the informal notes which I wrote for these talks and have been prepared in collaboration with Paulsen. It has been a true joint effort in all the best senses and it is doubtful that these notes could have reached fruition without his efforts.

Let me make one last remark in this unusual preface. Although much that is in these notes is not new, but a reformulation of known results, I believe that this framework can enable operator theory to encompass the case of several variables. This will require the efforts of many people and considerable time. The notes have been written as a kind of invitation to others to join this effort. Thus, in the final chapter we describe some preliminary results and possibilities for the module theoretic approach to multi-variate operator theory.

Ronald G. Douglas
Stony Brook, N.Y.
September, 1987

Introduction

Operator theory, which has been studied for about one hundred years, developed largely from the theory of differential and integral equations stemming from nineteenth century physics. About the time that the spectral theorem, compact operators, and the beginnings of functional analysis were understood, twentieth century physics in the form of quantum theory came along to raise new questions and to fundamentally enlarge the scope of operator theory. In particular, quantum theory made it clear that algebras of operators were also important and worthy of study. Even so, for the most part, the main interest continued to be in self-adjoint phenomena, whether for one operator or for an algebra of operators.

A change occurred in the fifties and sixties, however, when researchers began considering non-selfadjoint operators. One line of development focused on the study of shift operators and began with the work of Beurling [26], Lax [77], Halmos [68] and Helson and Lowedenslager [72] and culminated in the model theory of Sz-Nagy and Foiaş [101] for contraction operators. A parallel development of this theory occurred in the work of Livsitz [78] and Brodskii [29] in the Soviet Union as well as in the work of deBranges [28] in the United States. These three lines of development began independently but converged in the late sixties. About the same time engineers and scientists working in systems theory [32] and [33] were led to study similar problems for quite practical reasons. Ultimately a very deep, elegant, and useful theory was crafted from all of this work.

Almost from the beginning, researchers wondered about an analogous theory for a pair or for an n-tuple of commuting operators. Attempts at extending the theorem of Beurling which characterized the invariant subspaces for the shifts for the Hardy space of the polydisk were largely unsuccessful, although many interesting results were obtained [10], [94]. Moreover, no positive results were obtained which led toward a multivariate version of the model theory of Sz-Nagy and Foiaş. The problem is completely analogous to the situation in complex variables. Namely, most of the deep theorems for functions of one complex variable are false for functions of several complex variables. A naive attempt to go from the case of one to several complex variables results only in a profusion of counterexamples. What is necessary is that one start all over with new problems and new techniques, and the new techniques are more sophisticated and algebraic. This has resulted in a very elegant and useful theory for functions of several complex variables, but one that is rather different in flavor from that for functions of one complex variable. However, the new theory often casts new light on the old.

It is not just coincidental that the same situation exists in operator theory. In fact, it is not unreasonable to view the part of operator theory we are considering as "non-commutative complex variables." This is due, we believe, not just to the fact that we have made algebras of holomorphic functions central to the discussion. Rather we believe that this reflects the basic underlying structure theory for non-selfadjoint operators and algebras of operators. Evidence for this can be found in the study of hyponormal operators as well as in spectral theory and, we believe, in the approach taken in these notes. We do not claim, however, that this is the only point of view or yoga possible, only a very important one. Before beginning these notes we offer a brief description of the contents.

In Chapter 1 we study bounded and contractive Hilbert modules over the algebra C(X) of complex-valued continuous functions on a compact metrizable space X. We obtain a complete description of the modules in this case by showing that spectral multiplicity theory can be recast succinctly in this framework. Moreover, we provide a sketch of a powerful method for obtaining these results. In Chapter 2 we use the results obtained for C(X) as a guide to the study of Hilbert modules over more general function algebras concentrating in this chapter on the disk algebra. A class of "good modules", the so-called Šilov modules, is introduced. In Chapter 3, it is shown how dilation theory can be reformulated in terms of Šilov module resolutions. In all three chapters examples are provided and it is shown how this approach unifies results and problems from several disparate parts of operator theory.

The model theory of Sz-Nagy and Foiaş leads to an analysis of Šilov resolutions for contractive Hilbert modules for the disk algebra. This is described in Chapter 6. In the earlier Chapter 4, it is shown how the lifting theorem of Sz-Nagy and Foiaş can be reformulated in terms of a concept related to the algebraic concept of projectivity for modules. The derivation of results presented here is due to Foiaş and the first author and gives a new and different proof of the lifting theorem.

One of the most basic techniques in operator theory is that of localization to points of the spectrum. This is one source of the importance of functions of complex variables in the subject. We introduce localization in this context via the notion of module tensor product in Chapter 5. Applications of this technique are given in the final Chapter 6. These results, which are sometimes only sketched, include the derivation of the characteristic operator function and canonical models for contractive modules for the disk algebra, similarity invariants for the submodules of the Hardy module for the polydisk which have a zero-dimensional zero variety, and a study of the submodules corresponding to principal ideals of the polynomial algebra in several variables. We conclude by introducing the notion of a "locally free" Hilbert module via the localization sheaf. Finally, we sketch the role of Hermitian algebraic geometry in all of this and, in particular, show how the complex geometric methods introduced into operator theory by M. Cowen and the first author [38] extend to this context.

The results are only a beginning but we have high hopes for the future development of the subject.

1 The case of C(X)

In this chapter we establish the basic terminology and definitions for this study and introduce some standard examples. We do this while analyzing the Hilbert modules for the basic function algebra C(X) of continuous functions on a space X. We show that this analysis is equivalent to spectral multiplicity theory. The spectral theory enables one to localize the module action to the maximal ideals of C(X) and the multiplicity function then determines the Hilbert module up to either unitary equivalence or similarity. These results provide a framework for the study of Hilbert modules for general function algebras.

Let X be a compact, separable, metric space and let C(X) denote the algebra of all continuous complex-valued functions on X. We shall always regard C(X) as a Banach algebra equipped with the supremum norm,

$$\| f \| = \sup \{ | f(x) | : x \varepsilon X\}$$

for f in C(X).

A <u>function algebra on X</u> is a closed subalgebra of C(X), which contains the constant functions and separates points of X, that is, for every $x \neq y$ there is an f in the algebra with $f(x) \neq f(y)$.

We shall always use $\mathbb{C}$ to denote the complex plane and $\mathbb{D}$ to denote the open unit disk in $\mathbb{C}$. Some of the function algebras which we shall be interested in are listed below.

Example 1.1 a) Let $\mathbb{D}^-$ denote the closed unit disk. The disk algebra $A(\mathbb{D})$ is the closure in $C(\mathbb{D}^-)$ of the polynomials in the coordinate function z. The functions in $A(\mathbb{D})$ can be characterized as the functions in $C(\mathbb{D}^-)$ which are holomorphic on $\mathbb{D}$.

b) More generally, if $\mathbb{D}^n$ denotes the open polydisk in $\mathbb{C}^n$, then the polydisk algebra $A(\mathbb{D}^n)$ is the closure in $C(\mathbb{D}^{n-})$ of the polynomials in the coordinate functions $z_1, \ldots, z_n$.

c) Let $\mathbb{B}^n$ denote the closed unit ball in $\mathbb{C}^n$. The ball algebra $A(\mathbb{B}^n)$ is the closure in $C(\mathbb{B}^n)$ of the polynomials in the coordinate functions $z_1, \ldots, z_n$.

d) Let X be a compact set in $\mathbb{C}^n$. We let $\mathrm{Rat}(X)$ denote the algebra of rational functions, that is, quotients of polynomials with poles outside X, and let $R(X)$ denote the closure in $C(X)$ of $\mathrm{Rat}(X)$. By Runge's theorem, we have that $A(\mathbb{D}) = R(\mathbb{D}^-)$.

As the above examples indicate, we are primarily interested in "natural" function algebras which generally arise as algebras of holomorphic functions.

Definition 1.2. Let A be a function algebra, and let $\mathcal{H}$ be a Hilbert space. We say that $\mathcal{H}$ is a module over A provided that $\mathcal{H}$ is equipped with a mapping $A \times \mathcal{H} \to \mathcal{H}$, which we denote $(f,h) \to f\cdot h$, satisfying:

i) $1\cdot h = h$,

ii) $(fg)\cdot h = f\cdot(g\cdot h)$,

iii) $(f+g)\cdot h = f\cdot h + g\cdot h$,

iv) $f\cdot(\alpha h + \beta k) = \alpha\,(f\cdot h) + \beta(f\cdot k)$,

for every f, g in A, h, k in $\mathcal{H}$, and α, β in $\mathbb{C}$.

We call $\mathcal{H}$ a Hilbert module over A if, in addition, the mapping $A \times \mathcal{H} \to \mathcal{H}$ is separately continuous in each variable.

For f in A, we let $T_f: \mathcal{H} \to \mathcal{H}$ denote the linear map $T_f(h) = f\cdot h$. If $\mathcal{H}$ is a Hilbert module over A, then by the continuity in the second variable we have that T_f is bounded.

Proposition 1.3. Let $\mathcal{H}$ be a Hilbert module over A. Then there exists a constant K such that $\|T_f\| \leq K\|f\|$ for all f in A.

Proof. By the continuity in the first variable for each h in $\mathcal{H}$ there is a constant K_h such that $\|T_f h\| \leq K_h\|f\|$. Now apply the principle of uniform boundedness. □

Definition 1.4. Let $\mathcal{H}$ be a Hilbert module over A. Then the module bound of $\mathcal{H}$, is

$$K_A(\mathcal{H}) = \inf\{ K: ||T_f|| \leq K||f|| \text{ for all f in A}\}.$$

We call $\mathcal{H}$ contractive if $K_A(\mathcal{H}) = 1$.

There is an equivalent formulation of these notions in terms of algebraic representations. Given a Hilbert space $\mathcal{H}$, let $\mathcal{L}(\mathcal{H})$ denote the algebra of bounded, linear operators on $\mathcal{H}$.

The following is immediate.

Proposition 1.5. If $\mathcal{H}$ is a Hilbert module over A, then the map $\sigma: A \to \mathcal{L}(\mathcal{H})$ given by $\sigma(f) = T_f$ is a unital algebra homomorphism, σ is bounded, and $K_{\mathcal{H}} = ||\sigma||$.

We now present some examples of Hilbert modules.

Example 1.6. a) Let X be a separable metric space and let μ be a positive, Borel measure on X. Then the Hilbert space of square integrable functions $L^2(X,\mu)$ is a contractive Hilbert module for C(X) with the module action given by pointwise multiplication, that is, $(f\cdot h)(x) = f(x)\, h(x)$ for all f in $C(X)$ and h in $L^2(X,\mu)$.

b) Let N be a normal operator on a Hilbert space $\mathcal{H}$ and let X be the spectrum $\sigma(N)$ of N. By the spectral theorem for normal operators , for every f in $C(X)$ there is an operator $f(N)$ on $\mathcal{H}$. If we set $f\cdot h = f(N)h$ for f in $C(X)$ and h in $\mathcal{H}$, then $\mathcal{H}$ becomes a contractive Hilbert module over $C(X)$.

c) Let $\mathbb{T}$ be the unit circle in the complex plane, and let $L^2(\mathbb{T})$ be the Hilbert space of square integrable functions, with respect to arc-length measure. By the <u>Hardy space</u> $H^2(\mathbb{D})$ we mean the closed subspace of $L^2(\mathbb{T})$ spanned by the non-negative powers of the coordinate function z. If we set $(f\cdot h)(z) = f(z)h(z)$, for f in $A(\mathbb{D})$ and h in $H^2(\mathbb{D})$, then $f\cdot h$ is in $H^2(\mathbb{D})$. With this action $H^2(\mathbb{D})$ becomes a contractive Hilbert module over $A(\mathbb{D})$, which we call the <u>Hardy module.</u>

<u>d) The Bergman space</u> $B^2(\mathbb{D})$ is the closed subspace of the Hilbert space $L^2(\mathbb{D})$ of square integrable functions on $\mathbb{D}$ with respect to area measure which is spanned by the non-negative powers of z. Again, under pointwise multiplication, $B^2(\mathbb{D})$ becomes a contractive Hilbert module over $A(\mathbb{D})$, which we call the <u>Bergman module.</u>

e) Let A be any function algebra and let x be a point in its maximal ideal space. Then the space of complex numbers becomes a contractive Hilbert module, which we denote by $\mathbb{C}_x$, under the module action

$$f\cdot\lambda = f(x)\lambda.$$

f) Let A be any function algebra and let x and y be distinct points in its maximal ideal space, and let α be in $\mathbb{C}$. If we set,

$$T_f = \begin{pmatrix} f(x) & \alpha(f(x) - f(y)) \\ 0 & f(y) \end{pmatrix}$$

then $\mathbb{C}^2$ becomes a Hilbert module under the action $f\cdot h = T_f h$ for h in $\mathbb{C}^2$.

g) Let A be a function algebra and x a point in its maximal ideal space. A bounded linear functional $\delta : A \to \mathbb{C}$ is called an x-derivation if $\delta(fg) = f(x)\,\delta(g) + g(x)\,\delta(f)$. If we set

$$T_f = \begin{pmatrix} f(x) & \delta(f) \\ 0 & f(x) \end{pmatrix}$$

then $\mathbb{C}^2$ becomes a bounded Hilbert module under the action $f \cdot h = T_f h$ for h in $\mathbb{C}^2$.

For A = A($\mathbb{D}$), the maximal ideal space is $\mathbb{D}^-$, and the only points for which x-derivations exist are those in $\mathbb{D}$. The x-derivations are exactly the scalar multiples of the map $f \to f'(x)$. Similarly, if A = A($\mathbb{D}^n$), then x-derivations exist for x in $\mathbb{D}^n$ and correspond to scalar multiples of the analytic directional derivatives evaluated at x.

For these reasons, the set of x-derivations can be thought of as the tangent space at x of the algebra and the set of all derivations as the tangent space.

Because of the connection between x-derivations and derivatives, determining the module bound for the above action is intimately connected with the analytic structure of the algebra A. See [79] and [91] for some results along these lines.

Analyzing the three-dimensional Hilbert modules for a function algebra would require a knowledge of second order derivatives and their norms and so becomes very complicated.

The Hardy space and Bergman space are isomorphic as Hilbert spaces, but most studies of these spaces are really concerned with how they differ as A($\mathbb{D}$) modules.

<u>Definition 1.7.</u> Let $\mathcal{H}$ be a Hilbert module over A, and let $\mathcal{M}$ be a closed subspace of $\mathcal{H}$. Then $\mathcal{M}$ is a <u>submodule</u> of $\mathcal{H}$ for A, provided that $f \cdot h$ is in $\mathcal{M}$ for every f in A and h in $\mathcal{M}$.

Note that if $\mathcal{H}$ is a Hilbert module over A, and $\mathcal{M}$ is a submodule, then $\mathcal{M}$ is also a Hilbert module. Moreover, if $\mathcal{H}$ is a Hilbert module, then $\mathcal{M}$ is a Hilbert module, with $K_A(\mathcal{M}) \leq K_A(\mathcal{H})$.

For example 1.6b), $\mathcal{M}$ is a Hilbert submodule for C(X) if and only if $\mathcal{M}$ is a <u>reducing</u> subspace for N, that is, if and only if $N\mathcal{M} \subseteq \mathcal{M}$ and $N^*\mathcal{M} \subseteq \mathcal{M}$.

If we make $L^2(\mathbb{T})$ a Hilbert module over $A(\mathbb{D})$ via pointwise multiplication, then the Hardy space is a Hilbert submodule.

<u>Definition 1.8.</u> Let $\mathcal{H}$ and $\mathcal{K}$ be modules over A. Then a <u>module map</u> $X:\mathcal{H} \to \mathcal{K}$ is a bounded, linear map satisfying $X(f \cdot h) = f \cdot (Xh)$ for all f in A, and h in $\mathcal{H}$. Two Hilbert modules $\mathcal{H}$ and $\mathcal{K}$ are <u>similar</u> if there is an invertible module map from $\mathcal{H}$ onto $\mathcal{K}$, and are said to be <u>isomorphic</u> if there is a module map from $\mathcal{H}$ onto $\mathcal{K}$ which is a unitary.

We leave it to the reader to verify that every one-dimensional Hilbert module over A is isomorphic to the module $\mathbb{C}_x$ of example 1.6e), for some x in the maximal ideal space of A. Thus, every one-dimensional Hilbert module is necessarily contractive.

A more detailed analysis shows that every two-dimensional Hilbert module over A is isomorphic to one of the modules of example 1.6f) or of example 1.6g).

Moreover, if we fix x and y in example 1.6f), then for any two values of α, say α and α', the corresponding modules are always similar but are isomorphic if and only if $|\alpha| = |\alpha'|$.

For the remainder of this chapter, we focus on Hilbert modules over C(X). The structure that we exhibit in this case, to a large extent, frames our study of Hilbert modules over more general function algebras. In particular, we will be concerned with answering the following questions:

1. Can we give models for all contractive Hilbert modules over C(X)?

2. When are two models isomorphic and when are they similar?

3. Can we give models for all Hilbert modules over C(X)? In particular, is every Hilbert module over C(X) similar to a contractive module?

4. When are two models for Hilbert modules for C(X) similar?

5. If $\mathcal{H}$ and $\mathcal{K}$ are Hilbert modules for C(X), can we characterize the module maps?

The answers to these questions can be found, in some form, in most advanced texts on operator theory. To a large extent, they are translations of spectral multiplicity theory, or the reduction theory of commutative von Neumann algebras.

As we shall see below, every Hilbert module over C(X) is similar to a contractive module. Furthermore, contractive Hilbert modules over C(X) are similar if and only if they are isomorphic. Thus, to completely answer these questions, it will be sufficient to answer questions 1 and 2 and to characterize the module maps for contractive Hilbert modules over C(X).

Theorem 1.9. If $\mathcal{H}$ is a Hilbert module over C(X), then $\mathcal{H}$ is similar to a contractive Hilbert module over C(X).

Proof. Let $\mathcal{U}$ denote the unitary group of C(X) and let $\ell^\infty(\mathcal{U})$ be the bounded, measurable functions on $\mathcal{U}$.

For φ in $\ell^\infty(\mathcal{U})$ and u in $\mathcal{U}$, we set $\varphi_u(v) = \varphi(uv)$ and define $L_u : \ell^\infty(\mathcal{U})' \to \ell^\infty(\mathcal{U})'$ via $L_u\ (\delta)(\varphi) = \delta(\varphi_u)$ where δ is in $\ell^\infty(\mathcal{U})'$ and $\ell^\infty(\mathcal{U})'$ denotes the Banach space dual of $\ell^\infty(\mathcal{U})$.

By the Markov-Kakutani fixed point theorem [60], there is an m in $\ell^\infty(\mathcal{U})'$ such that $||m|| = 1$, $m(1) = 1$, m is positive, and $L_u(m) = m$ for all u, that is, $m(\varphi) = m\ (\varphi_u)$.

We use m to construct an equivalent inner-product on $\mathcal{H}$ such that $\mathcal{H}$ is a contractive Hilbert module in this new norm with respect to the old module action. The identity map from $\mathcal{H}$ with its old norm to $\mathcal{H}$ with its new norm is the module similarity.

For each x,y in $\mathcal{H}$, define $\varphi_{x,y}$ in $\ell^\infty(\mathcal{U})$ via $\varphi_{x,y}(u) = \langle u \cdot x,\ u \cdot y\rangle$. We set $\langle x,y\rangle_1 = m(\varphi_{x,y})$. It is routine to verify that this is an inner-product. The module action is contractive in this inner product because $\langle u \cdot x,\ u \cdot y\rangle_1 = \langle x,y\rangle_1$ so $T_u x = u \cdot x$ is a unitary in the new inner-product.

Finally, the inner-products are equivalent because,

$$||x||^2/K^2_{\mathcal{H}} \le \varphi_{x,x}(u) \le K^2_{\mathcal{H}}\ ||x||^2$$

and hence

$$||x||^2 / K^2_{\mathcal{H}} \le m\ (\varphi_{x,x}) = ||x||^2_1 \le K^2_{\mathcal{H}}\ ||x||^2. \quad \square$$

Theorem 1.10. If $\mathcal{H}_1$ and $\mathcal{H}_2$ are contractive Hilbert modules over C(X) which are similar, then they are isomorphic.

Proof. Let S denote the similarity. Since for each f in C(X), the operator T^i_f defined by $T^i_f h_i = f \cdot h_i$ for h_i in $\mathcal{H}_i$ is a normal operator for i = 1, 2, and $ST^1_f = T^2_f S$, we have that $T^1_f S^* = S^* T^2_f$. Hence, $U = S(S^*S)^{-1/2}$ is a unitary and $UT^1_f = T^2_f U$

for all f in C(X). □

Note that we can combine Theorems 1.9 and 1.10 to show that a Hilbert module over C(X) is similar to a unique isomorphism class of contractive Hilbert modules over C(X). Moreover, two Hilbert modules over C(X) are similar if and only if they determine the same contractive isomorphism class.

We begin our study of the first question by recalling the most familiar example of models. We assume throughout that X is a separable compact, metric space. If $\mathcal{H}$ is a Hilbert module over C(X), then we call a vector h in $\mathcal{H}$ cyclic, provided that $C(X)\cdot h = \{f\cdot h \mid f \in C(X)\}$ is dense in $\mathcal{H}$. We call $\mathcal{H}$ a cyclic Hilbert module if it has a cyclic vector. The following is well-known.

Lemma 1.11. Let $\mathcal{H}$ be a cyclic contractive Hilbert module over C(X). Then there is regular Borel measure ν on X with $\nu(X) = 1$, such that $\mathcal{H}$ and $L^2(X,\nu)$ are isomorphic Hilbert modules over C(X).

Proof. Let < , > denote the inner product on $\mathcal{H}$ and let h be a unit cyclic vector. By the Riesz representation theorem, there is a measure ν, as above, such that

$$<f\cdot h, h> = \int_X f\cdot d\nu$$

for all f in C(X). It is easily checked that setting $U(f\cdot 1) = f\cdot h$ extends to a unitary module map from $L^2(X,\nu)$ onto $\mathcal{H}$. □

Let Y be a topological space, φ: Y → X a continuous map and ν a Borel measure on Y. Then $L^2(Y,\nu)$ becomes a contractive module over C(X) under the action $(f\cdot h)(y) = f(\varphi(y))h(y)$ for f in C(X), and h in $L^2(Y,\nu)$. This is a special case of a general construction. If A_1 and A_2 are function algebras, $\chi: A_1 \to A_2$ is a bounded algebra homomorphism and $\mathcal{M}$ is a Hilbert module for A_2, then we can define then Hilbert module $\chi^*\mathcal{M}$ for A_1 such that $f\cdot h = \chi(f)\cdot h$. Moreover $\chi^*\mathcal{M}$ is contractive if both $\mathcal{M}$ and χ are contractive.

<u>Theorem</u> <u>1.12</u> Let $\mathcal{H}$ be a separable, contractive Hilbert module over C(X). Then there is a locally compact separable metric space Y, regular Borel measure ν on Y, $\nu(Y) = 1$, and a continuous function $\varphi: Y \to X$ such that $L^2(Y,\nu)$ and $\mathcal{H}$ are unitarily equivalent Hilbert modules over C(X).

<u>Proof</u>. Choose a countable family $\{h_n\}$ of unit vectors in $\mathcal{H}$ such that the spaces clos $\{C(X)\cdot h_n\} = \mathcal{H}_n$ are pairwise orthogonal and span $\mathcal{H}$. Apply Lemma 1.11 to each of these submodules $\mathcal{H}_n$ to obtain a measure ν_n, and let Y be the countable disjoint union of copies of X. Define ν on Y to be $2^{-n}\nu_n$ on the nth copy of X, and define $\varphi: Y \to X$ to be the identity map on each copy of X. □

Unfortunately, although the above models are quite simple, determining when two models are unitarily equivalent and determining the module maps between them is not as easy as for another set of models. These models are known as <u>direct integrals</u> and arise from fields of Hilbert spaces.

Let Y be a separable metric space and let ν be a σ-finite Borel measure on Y. We say that a subset E of Y is <u>ν-measurable</u> if E = B ∪ N where B is Borel and N is a subset of a Borel set N′ with $\nu(N') = 0$.

Suppose that for each y in Y we are given a separable Hilbert space $\mathcal{H}_y$, then a mapping $h: Y \to \bigcup_{y \in Y} \mathcal{H}_y$ such that h(y) is in $\mathcal{H}_y$ for each y in Y is called a <u>vector</u> <u>field</u>. The vector fields clearly form a vector space which we denote by $\mathcal{F}$.

<u>Definition</u> <u>1.13</u>. We say that $(\{\mathcal{H}_y\}_{y\varepsilon Y}, \Gamma)$ is a <u>ν-measurable</u> <u>field</u> <u>of</u> <u>Hilbert</u> <u>spaces</u> provided that Γ is a subspace of $\mathcal{F}$ such that:

i) for every h in Γ, the function $y \to \|h(y)\|$ is ν-measurable,

ii) if k is in $\mathcal{F}$ and for every h in Γ, the function $y \to < k(y), h(y) >$ is ν-measurable, then k is in Γ.

iii) there exists a sequence $\{h_n\}$ in Γ such that $\{h_n(y)\}$ is total in $\mathcal{H}_y$ for each y in Y.

A sequence $\{h_n\}$ satisfying iii) is called a <u>fundamental</u> <u>sequence</u> for Γ.

The <u>direct</u> <u>integral</u> $\int_Y \oplus \mathcal{H}_y \, d\nu$ is defined to be the set of equivalence classes of functions in Γ such that $\int_Y \|h(y)\|^2 d\nu$ is finite, where two vector fields are equivalent if they are equal ν-almost everywhere.

It is straightforward to verify that $\int_Y \oplus \mathcal{H}_y \, d\nu$ is a Hilbert space, with inner product $<h,k> = \int_Y <h(y),k(y)> d\nu$. Note that by i), the function $y \to <h(y),k(y)>$ is ν-measurable.

Now let X be a separable compact, metric space, ν a measure on X and $(\{\mathcal{H}_y\}_{y\varepsilon Y}, \Gamma)$ a ν-measurable field of Hilbert spaces. Then the direct integral becomes a contractive Hilbert module over C(X) in a natural fashion. If f is in C(X) and h is in Γ, then $f(x)\cdot h(x)$ is in Γ by definition 1.13ii), and so we may define $f\cdot h$ for h in $\int_Y \oplus \mathcal{H}_x \, d\nu$ via pointwise multiplication. By a theorem of Dixmier [46] this construction yields a model for every contractive Hilbert module over C(X).

We shall give an alternative proof of this fact using the theory of <u>disintegration of measures.</u> Let X and Y be locally compact separable metric spaces, and let P(Y) denote the set of positive regular Borel measures on Y. Suppose we are given ν in P(Y) and a Borel measurable map $\varphi: Y \to X$. If we define the measure μ on X via $\mu(E) = \nu(\varphi^{-1}(E))$, then μ is in P(X).

<u>Definition 1.14</u>. A <u>disintegration of ν with respect to φ</u> is a map from X into P(Y), $x \to \nu_x$ satisfying:

i) $\nu_x(Y \backslash \varphi^{-1}(\{x\})) = 0$ μ-almost everywhere,

ii) for every Borel set E of Y, the function $x \to \nu_x(E)$ is μ-measurable,

and

iii) $\nu(E) = \int_X \nu_x(E)\, d\mu$.

The existence of disintegration of measures under the above hypotheses is fairly well-known [6].

We now wish to use the disintegration of measures to construct a μ-measurable field of Hilbert spaces over X such that the model for a separable contractive Hilbert module over C(X) given in Theorem 1.12 as $L^2(Y,\nu)$ is unitarily equivalent to the C(X)-module, $\int_X \oplus \mathcal{H}_x\, d\mu$.

Suppose we are given a family of Hilbert spaces $\{\mathcal{H}_x\}_{x \in X}$ and a sequence of vector fields $\{h_n\}$ such that $\{h_n(x)\}$ is total in $\mathcal{H}_x$ for each x in X and such that $x \to \langle h_i(x), h_j(x) \rangle$ is μ-measurable for all i, j. If we let Γ denote the set of all vector fields h such that $x \to \langle h(x), h_j(x) \rangle$ is μ-measurable for all j, then it is easy to see that $(\{\mathcal{H}_x\}_{x \in X}, \Gamma)$ is a μ-measurable field of Hilbert spaces with fundamental sequence $\{h_n\}$. We call this the field <u>induced</u> by $\{h_n\}$.

Now let φ: $Y \to X$ and let ν be the measure in $P(Y)$ given by Theorem 1.12. Set $\mu(E) = \nu(\varphi^{-1}(E))$ and let $x \to \nu_x$ be a disintegration of ν with respect to φ, which may be chosen so that $\nu_x(Y) = 1$. If we let $\mathcal{H}_x = L^2(Y,\nu_x)$, then, since ν_x is a regular Borel measure and $\nu_x(Y)=1$, the space of continuous functions on Y which vanish at infinity $C_0(Y)$, is dense in $\mathcal{H}_x$. Choose a countable dense set $\{g_n\}$ of $C_0(Y)$, and let $\{h_n\}$ be the vector fields defined such that $h_n(x)$ is g_n viewed as a function in $L^2(Y,\nu_x)$. It is clear by ii) of the above definition that $x \to \langle h_i(x),h_j(x)\rangle = \int_X g_i \overline{g}_j \, d\nu_x$ is μ-measurable. Let Γ denote the field induced by these $\{h_n\}$.

<u>Theorem</u> <u>1.15.</u> Let Y, X be separable metric spaces, with X compact and Y locally compact, let ν be in $P(Y)$, let $\varphi: Y \to X$ be continuous, and let $(\{L^2(Y,\nu_x)\}, \Gamma)$ be the μ-measurable field of Hilbert spaces constructed above. Then the contractive Hilbert modules over $C(X)$ given by $L^2(Y,\nu)$ and $\int \oplus \mathcal{H}_x \, d\mu$ are isomorphic.

<u>Proof</u>. Let $C_0(Y)$ denote the space of continous functions on Y which vanish at infinity. A function g in $C_0(Y)$ is in $L^2(Y,\nu)$ and $L^2(Y,\nu_x)$ for all x. We let U_g be the vector field such that $(U_g)(x)$ is g as a function in $L^2(Y,\nu_x)$. Note that,

$$\|U_g\|^2 = \int_X \|(U_g)(x)\|^2 \, d\mu = \int_X \left[\int_Y |g(y)|^2 d\nu_x \right] d\mu = \int_Y |g(y)|^2 d\nu .$$

Thus, the map $g \to U_g$ is an isometry of $C_0(Y)$, regarded as a subspace of $L^2(Y,\nu)$, into $\int_X \oplus \mathcal{H}_x d\mu$. Since $C_0(Y)$ is dense in $L^2(Y,\nu)$ the map U can be extended to an isometry of $L^2(Y,\nu)$ into the direct integral, and we still denote this extension by U.

To see that U is onto, recall that we have a pre-assigned dense set $\{g_n\}$ of $C_0(Y)$ and corresponding vector fields $\{h_n\}$ which define Γ. But, by the definition, we see that $h_n = Ug_n$. Thus, any vector field orthogonal to the range of U, must be orthogonal to each h_n and hence 0.

Finally, to see that U is a C(X)-module map, note that for g in $C_0(Y)$, $(f \cdot g)(y) = f(\varphi(y))g(y) = f(x)g(y)$ ν_x-almost everywhere. Thus, $U(f \cdot g)(x) = f(x)(Ug)(x)$, which completes the proof. □

The above proof is only a slight modification of [6 ,Theorem 4].

<u>Corollary 1.16.</u> Let X be a separable, compact metric space and let $\mathcal{H}$ be a separable, contractive Hilbert module over C(X). Then there exists a σ-finite Borel measure μ on X and a μ-measurable field of Hilbert spaces, such that $\mathcal{H}$ and the direct integral are isomorphic Hilbert modules on C(X).

We now turn our attention to determining when two direct integral models are isomorphic. Before doing this, it will be convenient to first describe the module maps between two direct integrals.

Let X be a separable, compact metric space and let ν and ν' be two σ-finite Borel measures on X. Let $\mu = \nu + \nu'$ so that by the Radon-Nikodym theorem $d\nu = f d\mu$, and $d\nu' = f' \, d\mu$.

Let $(\{\mathcal{H}_x\},\Gamma)$ and $(\{\mathcal{H}'_x\},\Gamma')$ be, respectively, ν-measurable and ν'-measurable fields of Hilbert spaces over X. If for every x in X, T_x is in $\mathcal{L}(\mathcal{H}_x, \mathcal{H}_x')$, then $x \to T_x$ is called a <u>field of continuous linear mappings.</u>

Definition 1.17. We say that a field of continuous linear mappings T is <u>(Γ,Γ')-measurable</u> if for every h in Γ the vector field $x \to T_x h_x$ is in Γ'. We say that the field is <u>bounded</u> if $f'(x)\|T_x\|^2/f(x)$ is $(\nu + \nu')$-essentially bounded, when we set $0/0 = 0$.

Lemma 1.18. Let T_x be (Γ,Γ')-measurable and assume that $f'(x)\ \|T_x\|^2/f(x)$ is $(\nu + \nu')$-essentially bounded with essential bound λ^2. Then T_x defines a bounded module map $T\colon \int_X \oplus \mathcal{H}_x\, d\nu \to \int_X \oplus \mathcal{H}'_x d\nu'$ with $\|T\| = \lambda$.

Definition 1.19. A bounded linear operator $T\colon \int_X \oplus \mathcal{H}_x d\nu \to \int_X \oplus \mathcal{H}'_x d\nu'$ is <u>decomposable</u> if it is given by a bounded (Γ,Γ')-measurable field as above. We say that T is the <u>integral</u> of this field.

Theorem 1.20. Let $(\{\mathcal{H}_x\},\ \Gamma)$, $(\{\mathcal{H}_x\},\ \Gamma')$ be, respectively, ν-measurable and ν'-measurable fields of Hilbert spaces on the separable, compact metric space X. Then a map $T\colon \int_X \oplus \mathcal{H}_x d\nu \to \int_X \oplus \mathcal{H}'_x d\nu'$ is a bounded Hilbert module map over C(X) if and only if it is decomposable.

Proof. See the proof of [46 ,Theorem II.2.1], where the $\nu = \nu'$ case is proved. □

Now that we have a clear picture of the module maps, it is possible to analyze when two of the models for contractive Hilbert modules over C(X) are module isomorphic.

Proposition 1.21. Let $(\{\mathcal{H}_y\}_{y\in Y},\Gamma)$ be a ν-measurable field of Hilbert spaces. Then the function $d\colon Y \to \{0,1,\ldots,\aleph_0\}$

defined by $d(y) = \dim(\mathcal{H}_y)$ is ν-measurable. Conversely, given a ν-measurable function $d(y)$ there exists a ν-measurable field of Hilbert spaces, with $d(y) = \dim(\mathcal{H}_y)$.

<u>Proof.</u> Let $\{h_n\}$ be a fundamental sequence for Γ and let $\mathcal{G}_n(y) = (\langle h_i(y), h_j(y)\rangle)$ be the $n \times n$ Grammian. Clearly, $\gamma_n(y) = \operatorname{rank} \mathcal{G}_n(y)$ is ν-measurable and hence, $d(y) = \sup_n \gamma_n(y)$ is ν-measurable. For the converse, let $Y_k = d^{-1}(k)$, and fix a separable Hilbert space $\mathcal{M}$ with orthonormal basis $\{e_k\}^{\infty}{}_{k=1}$. Let Γ be the set of all ν-measurable functions from Y to $\mathcal{M}$ such that $h(y)$ is in the span of $\{e_k\}^{n}{}_{k=1}$ for y in Y_n, and define $\{h_n\}$ by

$$h_n(y) = \begin{Bmatrix} e_n \text{ for } y \in \bigcup_{j=n}^{\infty} Y_j \\ 0 \text{ otherwise} \end{Bmatrix} .$$

It is easily checked that this is a ν-measurable field of Hilbert spaces, with $\mathcal{H}_y = \operatorname{span}\{e_1, \ldots, e_n\}$ for y in Y_n. □

Fix a separable Hilbert space $\mathcal{H}$, then by the <u>constant ν-measurable field with fiber $\mathcal{H}$</u>, we mean the field $(\{\mathcal{H}_y\}_{y \in Y}, \Gamma)$ where $\mathcal{H}_y = \mathcal{H}$ for all y, and Γ is the set of all ν-measurable maps. If we fix an orthonormal basis $\{e_n\}$ for $\mathcal{H}$, then setting $h_n(y) = e_n$ defines a fundamental sequence.

The direct integral of the constant ν-measurable field with fiber $\mathcal{H}$ is clearly identical to $L^2{}_{\mathcal{H}}(Y,\nu)$, the space of square integrable $\mathcal{H}$-valued functions on Y. If Y is a compact, metric space and we let $C(Y)$ act on $L^2{}_{\mathcal{H}}(Y,\nu)$ via pointwise multiplication, then we see by Theorem 1.18 that the operators on $L^2{}_{\mathcal{H}}(Y,\nu)$ which commute with this action are all given by the pointwise multiplication by essentially bounded ν-measurable $\mathcal{L}(\mathcal{H})$-valued functions.

Lemma 1.22. Let $(\{\mathcal{H}_y\}_{y\varepsilon Y},\Gamma)$ be a ν-measurable field of Hilbert spaces with $\dim(\mathcal{H}_y)$ constant, fix a Hilbert space $\mathcal{H}'$ of the same dimension and let $(\{\mathcal{H}'_y\}_{y\varepsilon Y},\Gamma')$ be the constant ν-measurable field with fiber $\mathcal{H}'$. Then there exists a bounded (Γ,Γ')-measurable field of unitaries $U_y: \mathcal{H}_y \to \mathcal{H}'$ such that $\Gamma' = \{U_y\ h(y): h\ \varepsilon\ \Gamma\ \}$.

Proof. Let $\{h_n\}$ be a fundamental sequence for Γ and let $d = \dim(\mathcal{H}_y)$. For $k < d + 1$ let $g_k(y)$ be the k-th element of the sequence obtained from applying the Gramm-Schmidt process to the (possibly linearly dependent) sequence $\{h_n(y)\}$. It is not difficult to see that the g_k's defined in this fashion are ν-measurable, and form a new fundamental sequence for Γ.

Fix an othonormal basis, $\{e_k\}$ for $\mathcal{H}'$ and define $U_y: \mathcal{H}_y \to \mathcal{H}'$ via $U_y g_k(y) = e_k$. It is not difficult to see that these maps have the desired properties, which concludes the proof. □

Theorem 1.23. Let X be a compact, separable metric space, let $\mathcal{H}$ be a separable, contractive Hilbert module over C(X), and fix Hilbert spaces $\mathcal{H}_n$, $n=1,2,\ldots,\aleph$ of dimension n. Then there is a Borel measure ν on X and disjoint subsets X_n of X, $n=1,2,\ldots,\aleph$ such that $\mathcal{H}$ is C(X)-module isomorphic to $\sum\oplus L^2_{\mathcal{H}_n}(X_n,\nu)$, where the module action on each $L^2_{\mathcal{H}_n}(X_n,\nu)$ is defined via pointwise multiplication.

Proof. Apply Lemma 1.20 and Proposition 1.19 to Theorem 1.14, with $X_n = \{x: \dim(\mathcal{H}_x) = n\}$. □

Theorem 1.24. Let X be a compact, separable metric space and let $(\{\mathcal{H}_x\}_{x\varepsilon X},\Gamma)$ and $(\{\mathcal{H}'_x\}_{x\varepsilon X},\Gamma')$ be, respectively, ν-measurable and ν'-measurable fields of Hilbert spaces. Then $\int_X\oplus\mathcal{H}_x d\nu$ and $\int_X\oplus\mathcal{H}'_x d\nu'$ are C(X)-module isomorphic if and only if

ν and ν' are mutually absolutely continuous and dim $(\mathcal{H}_x)=$ $\dim(\mathcal{H}'_x)$ ν-almost everywhere.

<u>Proof</u>. Let $X_n = \{x: \dim(\mathcal{H}_x) = n\}$, $X'_n = \{x: \dim(\mathcal{H}'_x) = n\}$, and fix Hilbert spaces $\mathcal{K}_n$, $\dim(\mathcal{K}_n) = n$, $n = 1,2,\ldots,\aleph_0$. By Theorem 1.23 we may assume that the spaces are $\sum \oplus L^2_{\mathcal{K}_n}(X_n,\nu)$ and $\sum\oplus L^2_{\mathcal{K}_n}(X'_n,\nu')$. If U defines a C(X)-module isomorphism, then by Theorem 1.18 U is defined by a field U_x. Similarly, U^{-1} is defined by a field V_x. By Definition 1.15, there is a set N, $(\nu+\nu')(N) = 0$ such that if $E \cap N = \emptyset$ and $\nu\wedge\nu'(E)=0$, then U_x and V_x are 0 for x in E. Thus, if $\nu(E) > 0$ but $\nu'(E) = 0$ then $U_x = 0$ which shows that U has a nontrivial kernel. Hence, ν is absolutely continous with respect to ν'. The converse statement follows similarly.

Now in order for U to be an isometry on $L^2_{\mathcal{K}_n}(X_n,\nu)$ we must have that $X_n \subseteq \bigcup_{j\geq n} X'_j$ ν-almost everywhere. Similarly, $X'_n \subseteq \bigcup_{j\geq n} X_j$ ν'-almost everywhere follows by considering U^{-1}. Thus, after removing a set N with $\nu(N) = \nu'(N) = 0$, we have that these containments are true everywhere, and so $X_n\backslash N = X'_n\backslash N$.

The remainder of the proof is routine. □

It is worthwhile to note that by the above result, the particular space of sections Γ plays no essential role in determining the direct integral up to isomorphism. For these reasons, it is often suppressed in what follows.

Before continuing we summarize the basic facts about Hilbert modules over C(X). First, every contractive Hilbert module is isomorphic to a direct integral module which is determined up to isomorphism by its multiplicity. Second, every Hilbert module is similar to a contractive Hilbert module and hence Hilbert modules are also determined up to similarity by multiplicity. For no other function algebra do we understand the basic module structure in this detail.

We close this chapter with one other module theoretic concept which will be useful. If $\mathcal{M}$ is a Hilbert module for A, then a set $\{h_\delta\}_{\delta\varepsilon\Gamma} \subseteq \mathcal{M}$ is called a <u>generating set</u> for $\mathcal{M}$ if finite linear sums of the form

$$\sum_i f_i \cdot h_{\delta_i}, \quad f_i \varepsilon A, \ \delta_i \varepsilon \Gamma$$

are dense in $\mathcal{M}$.

<u>Definition 1.25.</u> If $\mathcal{M}$ is a Hilbert module over A, then $\mathrm{rank}_A(\mathcal{M})$, the <u>rank of $\mathcal{M}$, over A</u>, is the minimum cardinality of a generating set for $\mathcal{M}$.

Thus, $\mathcal{M}$ is cyclic if and only if $\mathrm{rank}_A(\mathcal{M}) = 1$. For modules over C(X) it is easy to determine the rank from the models. The following is immediate.

<u>Proposition 1.26.</u> Let $\mathcal{H}=\int_X \oplus \mathcal{H}_x \, d\nu$ be a direct integral of Hilbert spaces. Then the rank of $\mathcal{H}$ as a C(X)-module is the ν-essential supremum of the dimension function dim $(\mathcal{H}_x)$.

<u>Proposition 1.27.</u> Let $\mathcal{H}$ be a bounded, Hilbert module over C(X) and let $\mathcal{K} \subseteq \mathcal{H}$ be a submodule, then $\mathrm{rank}_{C(X)}\mathcal{K} \leq \mathrm{rank}_{C(X)}\mathcal{H}$.

Although the notion of module rank is analogous to that of the dimension of a linear space, they differ in fundamental ways. For example, in [21], two Hilbert submodules $\mathcal{M}$ and $\mathcal{N}$, $\mathcal{M} \subseteq \mathcal{N}$, of the Bergman module $B^2(\mathbb{D})$ over $A(\mathbb{D})$ are exhibited such that the zero module action is induced on $\mathcal{N}/\mathcal{M}$ and $\dim_{\mathbb{C}}\mathcal{N}/\mathcal{M}$ is infinite. It follows that $\mathrm{rank}_{A(\mathbb{D})}\mathcal{N}$ is infinite. Therefore, in general, little can be said about the rank of submodules.

2 The disk algebra and Šilov modules

In the last chapter, we gave a rather complete description of the Hilbert modules for $C(X)$. In this chapter, we consider the analogous problem for the disk algebra $A(\mathbb{D})$. We shall see that for contractive Hilbert modules over $A(\mathbb{D})$, this problem is equivalent to describing all contraction operators. We begin by introducing a smaller class of modules, the Šilov modules, which can be thoroughly analyzed. In Chapter 3, we then show that the contractive Hilbert modules for $A(\mathbb{D})$ can be analyzed by considering their resolutions in terms of Šilov modules.

If T is an operator on the Hilbert space $\mathcal{H}$, then we can always make $\mathcal{H}$ into a module over the algebra $\mathbb{C}[z]$ of polynomials in z, by setting $p \cdot h = p(T)h$, for p in $\mathbb{C}[z]$ and h in $\mathcal{H}$. In general, this does not extend to make $\mathcal{H}$ into a (bounded) Hilbert module over $A(\mathbb{D})$. Clearly, this extends to make $\mathcal{H}$ into a Hilbert module over $A(\mathbb{D})$ if and only if there exists a constant K, such that

$$\|p(T)\| \leq K\|p(z)\|_{A(\mathbb{D})}$$

for all polynomials p in $\mathbb{C}[z]$, and the module norm $K_{\mathcal{H}}$ of $\mathcal{H}$ will be the least such constant.

<u>Definition</u> <u>2.1.</u> An operator T on a Hilbert space $\mathcal{H}$ is <u>polynomially</u> <u>bounded</u> if there exists a constant K such that $\|p(T)\| \leq K\|p(z)\|_{A(\mathbb{D})}$ for all polynomials p.

The following is immediate:

Proposition 2.2. There is a one-to-one correspondence between the Hilbert modules for $A(\mathbb{D})$ and polynomially bounded operators, where the Hilbert module $\mathcal{H}$ corresponds to the operator T defined by setting $Th=z\cdot h$, for h in the Hilbert module $\mathcal{H}$. Two Hilbert modules for $A(\mathbb{D})$ are similar if and only if the corresponding operators are similar, and isomorphic if and only if the corresponding operators are unitarily equivalent.

How do we determine which operators give rise to contractive Hilbert modules? Clearly, a necessary condition is that $||T|| \leq ||z|| = 1$, or that T be a contraction. That this condition is also sufficient is due to von Neumann [87]:

Theorem 2.3(von Neumann). If T is a contraction operator on a Hilbert space, then $\|p(T)\| \leq \|p(z)\|_{A(\mathbb{D})}$ for all polynomials p.

Consequently there is a one-to-one correspondence between contractive Hilbert modules for $A(\mathbb{D})$ and contraction operators. Thus the problem of classifying all contractive Hilbert modules for $A(\mathbb{D})$ up to isomorphism is the same as classifying all contraction operators up to unitary equivalence.

If an operator is similar to a contraction operator, then it is polynomially bounded. Whether or not all polynomially bounded operators arise in this manner is an unsolved problem due to Sz.-Nagy and Halmos [69] which in our setting becomes:

Problem 2.4. Is every Hilbert module for $A(\mathbb{D})$ similar to a contractive Hilbert module?

We saw in the last chapter that every Hilbert module over $C(X)$ is similar to a contractive Hilbert module over $C(X)$. However, for no other function algebra has this problem been solved either positively or negatively.

Problem 2.5. Does there exist a Hilbert module for some function algebra which is not similar to a contractive Hilbert module?

Also, in contrast to the case for $C(X)$ we have that contractive Hilbert modules for $A(\mathbb{D})$ which are similar need not be isomorphic, since similar contraction operators need not be unitarily equivalent. Consider for example the operators on $\mathbb{C}^2$ defined by the matrices

$$\begin{bmatrix} 0 & 1 \\ 0 & 0 \end{bmatrix} \quad \text{and} \quad \begin{bmatrix} 0 & \frac{1}{2} \\ 0 & 0 \end{bmatrix}.$$

Most "natural" function algebras not isomorphic to $C(X)$ can be easily shown to have contractive Hilbert modules which are similar but not isomorphic.

Problem 2.6. If A is a function algebra for which the relations of isomorphism and similarity coincide for contractive Hilbert modules, must A be isomorphic to $C(X)$ for some compact, Hausdorff space X?

We now turn our attention to $A(\mathbb{D}^n)$. If $T_1,\ldots,T_n$ are operators on $\mathcal{H}$ which pairwise commute, then $\mathcal{H}$ becomes a module over the algebra $\mathbb{C}[z_1,z_2,\ldots,z_n]$ of polynomials in n-variables $z_1,\ldots,z_n$ by setting $p\cdot h = p(T_1,\ldots,T_n)h$ for p in $\mathbb{C}[z_1,\ldots,z_n]$ and h in $\mathcal{H}$. For this action to make $\mathcal{H}$ a contractive Hilbert module over $A(\mathbb{D}^n)$ it is necessary that $\|T_i\| \leq 1$, $i = 1,2,\ldots,n$. When $n = 1$, we saw that this condition is also sufficient, a result due to von Neumann. When $n = 2$, this is also sufficient and follows from a result of Ando [11] which we return to in Chapter 4. However, for $n \geq 3$ this condition is not sufficient, as examples of Varopoulos [104] and Crabbe and Davie [41] show.

Thus, while the contractive Hilbert modules over $A(\mathbb{D}^2)$ are in one-to-one correspondence with the pairs of commuting contractions, we currently have little knowledge of how a general contractive Hilbert module over $A(\mathbb{D}^n)$ arises, $n \geq 3$. It is known that if the n-tuple of contraction operators $T_1, \ldots, T_n$ doubly commutes, that is, if $T_iT_j = T_jT_i$ and $T_iT_j^* = T_j^*T_i$ for $i \neq j$, then the n-tuple determines a contractive Hilbert module over $A(\mathbb{D}^n)$ with the action defined above [101] .

Problem 2.7. Give necessary and sufficient conditions for n commuting contractions to determine a contractive Hilbert module over $A(\mathbb{D}^n)$.

Although n commuting contractions do not necessarily yield a contractive Hilbert module over $A(\mathbb{D}^n)$, no examples are known where they do not yield a (bounded) Hilbert module. This has been studied fairly extensively ([47], [48], and [104]).

Problem 2.8. Does there exist a constant K_n such that if $T_1, \ldots, T_n$ are commuting contractions, then $\|p(T_1, \ldots, T_n)\| \leq K_n \|p(z_1, \ldots, z_n)\|_{A(\mathbb{D}^n)}$ for all polynomials p?

Results of Varopoulos [104] show that if such constants exist, then K_n tends to infinity as n tends to infinity.

In order to analyze the contractive Hilbert modules over $A(\mathbb{D})$, we first introduce a subclass, the Šilov modules which are fairly easy to analyze. We will later analyze arbitrary contractive Hilbert modules over $A(\mathbb{D})$ by resolving them using Šilov modules (Chapter 3).

Recall that if A is a function algebra and M_A is its maximal ideal space, then via the Gelfand transform we may regard A as a function algebra on M_A, that is, $A \subseteq C(M_A)$. The Šilov boundary for A, denoted ∂A, is the smallest closed subset of M_A on which each function in A achieves its maximum modulus. Thus, we may regard A as a function algebra on ∂A.

The existence of the Šilov boundary for a function algebra is a deep result due to Šilov. See [63] for a proof. For $A(\mathbb{D})$ the Šilov boundary is $\partial\mathbb{D} = \mathbb{T}$. For a compact set X in $\mathbb{C}$, the Šilov boundary of R(X) is ∂X.

Definition 2.9. Let $A \subseteq C(\partial A)$ be a function algebra and let $\mathcal{M}$ be a contractive Hilbert module over $C(\partial A)$. A closed subspace $\mathcal{S} \subseteq \mathcal{M}$ which is invariant for A is called a Šilov module over A. A Šilov module for A is reductive if it is invariant for $C(\partial A)$ and pure if no non-zero subspace of it is reductive.

It is easy to see that the direct sum of Šilov modules for A is a Šilov module for A and the direct sum will be a pure Šilov module if each summand is pure. Also, submodules of Šilov modules are Šilov, and submodules of pure Šilov modules are pure.

Proposition 2.10. Let $\mathcal{S}$ be a Šilov module for A. Then $\mathcal{S}$ has a unique decompositon as $\mathcal{S} = \mathcal{S}_p \oplus \mathcal{S}_r$ where $\mathcal{S}_p \subseteq \mathcal{S}$ and $\mathcal{S}_r \subseteq \mathcal{S}$ are, respectively, pure and reductive submodules of $\mathcal{S}$.

Proof. Let $\mathcal{S} \subseteq \mathcal{H}$, where $\mathcal{H}$ is a contractive Hilbert module for $C(\partial A)$, and $\mathcal{S}$ is a Hilbert submodule for A. It is not hard to see that $\mathcal{S}$ contains a maximal subspace $\mathcal{S}_r$ which is invariant for $C(\partial A)$. Moreover, since $\mathcal{S}_r^{\perp}$ is invariant for A, it follows that $\mathcal{S}_p = \mathcal{S} \cap \mathcal{S}_r^{\perp}$ is a Hilbert submodule for A and that

$\mathcal{S} = \mathcal{S}_p \oplus \mathcal{S}_r$. □

In the one-to-one correspondence between contractive Hilbert modules over $A(\mathbb{D})$ and contraction operators, we see that Šilov modules over $A(\mathbb{D})$ correspond to isometries. A Šilov module over $A(\mathbb{D})$ is reductive if it corresponds to a unitary and pure if it corresponds to an isometry with no unitary part, that is, a pure isometry.

Fix a Hilbert space $\mathcal{E}$ and let $L^2{}_{\mathcal{E}}(\mathbb{T})$ denote the Hilbert space of square-integrable $\mathcal{E}$-valued functions on $\mathbb{T}$ with respect to arc-length measure. Let $H^2{}_{\mathcal{E}}(\mathbb{D})$ denote the closed subspace of $L^2{}_{\mathcal{E}}(\mathbb{T})$ consisting of those functions whose negative Fourier coefficients vanish, that is, those f for which

$$\int_0^{2\pi} f(e^{it})\, e^{ikt}\, dt = 0,$$

for $k > 0$. The space $L^2{}_{\mathcal{E}}(\mathbb{T})$ is a contractive Hilbert module over $C(\mathbb{T})$ under pointwise multiplication and hence $H^2{}_{\mathcal{E}}(\mathbb{D})$ is a pure Šilov module for $A(\mathbb{D})$. The corresponding operator of multiplication by z is a unilateral shift of multiplicity equal to the dimension of $\mathcal{E}$. It is not hard to see that $H^2{}_{\mathcal{E}_1}(\mathbb{D})$ and $H^2{}_{\mathcal{E}_2}(\mathbb{D})$ are isomorphic as $A(\mathbb{D})$-modules if and only if $\dim(\mathcal{E}_1) = \dim(\mathcal{E}_2)$. These Šilov modules serve as models for the pure Šilov modules for $A(\mathbb{D})$, as we shall soon see.

The well-known theorem of von Neumann and Wold on the classification of isometries [86] and [106] can be reformulated to characterize Šilov modules for $A(\mathbb{D})$.

<u>Theorem 2.11(von Neumann-Wold).</u> Every Šilov module for $A(\mathbb{D})$ is module isomorphic to $H^2{}_{\mathcal{E}}(\mathbb{D}) \oplus \mathbf{R}$ where $\mathbf{R}$ is a reductive Šilov module. If Šilov modules $\mathcal{S}_i$ are isomorphic to $H^2{}_{\mathcal{E}_i}(\mathbb{D}) \oplus \mathbf{R}_i$, $i = 1, 2$, then $\mathcal{S}_1$ and $\mathcal{S}_2$ are similar if and only if $\dim(\mathcal{E}_1) = \dim(\mathcal{E}_2)$ and $\mathbf{R}_1$ and $\mathbf{R}_2$ are isomorphic. Moreover, two

Šilov modules for $A(\mathbb{D})$ are similar if and only if they are isomorphic.

<u>Proof.</u> The decomposition of a Šilov module as $H^2_{\mathcal{E}}(\mathbb{D})\oplus\mathbb{R}$ follows from the decomposition of an isometry as the direct sum of a pure isometry and a unitary .

Assume that $\mathcal{S}_i = H^2_{\mathcal{E}_i}(\mathbb{D})\oplus\mathbb{R}_i$, $i = 1,2$, and that $\Phi\colon \mathcal{S}_1\to\mathcal{S}_2$ is a similarity. Since $\mathbb{R}_i = \bigcap_{n\geq 1} z^n\mathcal{S}_i$, $i = 1,2$, and $\Phi(z^n\mathcal{S}_1) = z^n\mathcal{S}_2$, we see that the restriction of Φ to $\mathbb{R}_1$ is a similarity between $\mathbb{R}_1$ and $\mathbb{R}_2$, as $A(\mathbb{D})$-modules. But $\mathbb{R}_1$ and $\mathbb{R}_2$ are also $C(\mathbb{T})$-modules, and it is readily verified that any $A(\mathbb{D})$-module map between $\mathbb{R}_1$ and $\mathbb{R}_2$ is a $C(\mathbb{T})$-module map. Thus, $\mathbb{R}_1$ and $\mathbb{R}_2$ are similar as $C(\mathbb{T})$-modules and hence are isomorphic.

Since $\dim(\mathcal{E}_i) = \dim(\mathcal{S}_i\ominus z\mathcal{S}_i)$ and $\Phi(z\mathcal{S}_1) = z\mathcal{S}_2$, we have that $\dim(\mathcal{E}_1) = \dim(\mathcal{E}_2)$. Thus, if $\mathcal{S}_1$ and $\mathcal{S}_2$ are similar, then $\dim(\mathcal{E}_1) = \dim(\mathcal{E}_2)$ and $\mathbb{R}_1$ and $\mathbb{R}_2$ are similar.

Finally, if $\dim(\mathcal{E}_1) = \dim(\mathcal{E}_2)$ and $\mathbb{R}_1$ and $\mathbb{R}_2$ are isomorphic, then it is easily seen that $\mathcal{S}_1$ and $\mathcal{S}_2$ are isomorphic, from which the remaining claims of the theorem follow. □

Let $\mathcal{M}$ be a Hilbert module over A and let $\mathcal{H}$ be a subspace of $\mathcal{M}$. We let $A\cdot\mathcal{H}$ denote the vector space spanned by all products $f\cdot h$ for f in A and h in $\mathcal{H}$. The closure $[A\cdot\mathcal{H}]^-$ of $A\cdot\mathcal{H}$ will be a submodule of $\mathcal{M}$.

Now let $\mathcal{S}$ be a Šilov module for A so that $\mathcal{S}\subseteq\mathcal{M}$ where $\mathcal{M}$ is a contractive Hilbert module for $C(\partial A)$. Clearly, $[C(\partial A)\cdot\mathcal{S}]^-$ is the smallest submodule of $\mathcal{M}$ for $C(\partial A)$ that contains $\mathcal{S}$.

Definition 2.12. Let $\mathcal{S}$ be a Šilov module for A. If $\mathcal{M}$ is a contractive Hilbert module for $C(\partial A)$ which contains $\mathcal{S}$ and if $C(\partial A)\cdot\mathcal{S}$ is dense in $\mathcal{M}$, then $\mathcal{M}$ is called a minimal $C(\partial A)$-extension of $\mathcal{S}$.

Note that since $\overline{A}$ and A generate $C(\partial A)$ algebraically, $C(\partial A)\cdot\mathcal{S}$ is dense in $\mathcal{M}$ if and only if $\overline{A}\cdot\mathcal{S}$ is dense in $\mathcal{M}$.

Proposition 2.13. Let $\mathcal{S}$ be a Šilov module for A and let $\mathcal{M}$ be a minimal $C(\partial A)$-extension of $\mathcal{S}$. If $\mathcal{M}'$ is a contractive Hilbert module over $C(\partial A)$ and $V\colon \mathcal{S}\to\mathcal{M}'$ is an A-module isometry, then V extends uniquely to a $C(\partial A)$-module isometry $\tilde{V}\colon \mathcal{M}\to\mathcal{M}'$.

Proof. If $\tilde{V}$ exists, then necessarily $\tilde{V}(\overline{f}\cdot h) = \overline{f}\cdot Vh$ for h in $\mathcal{S}$ and f in A. It is easily checked that this formula extends linearly to yield an isometry since,

$$\|\textstyle\sum_i \overline{f}_i\cdot Vh_i\|^2 = \sum_{i,j} \langle \overline{f}_i\cdot Vh_i, \overline{f}_j Vh_j\rangle_{\mathcal{M}'}$$

$$= \textstyle\sum_{i,j}\langle f_j\cdot Vh_i, f_i\cdot Vh_j\rangle_{\mathcal{M}'} = \sum_{i,j}\langle f_j\cdot h_i, f_i\cdot h_j\rangle_{\mathcal{S}}$$

$$= \textstyle\sum_{i,j}\langle \overline{f}_i\cdot h_i, \overline{f}_i\cdot h_j\rangle_{\mathcal{M}} = \|\sum \overline{f}_i\cdot h_i\|^2, \text{ for any finite}$$

set, $f_1, \ldots, f_n$ in A and $h_1, \ldots h_n$ in $\mathcal{S}$. □

Corollary 2.14. Let $\mathcal{S}$ be a Šilov module for A. If $\mathcal{M}$ and $\mathcal{M}'$ are minimal $C(\partial A)$-extensions of $\mathcal{S}$, then they are isomorphic as $C(\partial A)$-modules via an isomorphism which is the identity on $\mathcal{S}$.

By the above corollary, we see that a minimal $C(\partial A)$-extension of a Šilov module is unique up to $C(\partial A)$-module isomorphism. The minimal $C(\partial A)$-extension is also useful for calculating rank.

<u>Proposition 2.15.</u> Let $\mathcal{S}$ be a Šilov module for A and let $\mathcal{K}$ be the minimal $C(\partial A)$-extension of $\mathcal{S}$.

Then $\text{rank}_A(\mathcal{S}) \geq \text{rank}_{C(\partial A)}(\mathcal{K})$.

<u>Proof.</u> Since $C(\partial A)\cdot\mathcal{S}$ is dense in $\mathcal{K}$, any generating set for $\mathcal{S}$ over A is a generating set for $\mathcal{K}$ over $C(\partial A)$.

For the Šilov module of the form $H^2_{\mathcal{E}}(\mathbb{D})\oplus\mathcal{R}$ over $A(\mathbb{D})$, it is clear that its minimal $C(\mathbb{T})$-extension is $L^2_{\mathcal{E}}(\mathbb{T})\oplus\mathcal{R}$. Also, $\text{rank}_{A(\mathbb{D})}(H^2_{\mathcal{E}}(\mathbb{D})) = \dim(\mathcal{E}) = \text{rank}_{C(\mathbb{T})}(L^2_{\mathcal{E}}(\mathbb{T}))$, and consequently by Theorem 2.11, $\text{rank}_{A(\mathbb{D})}(\mathcal{S}) = \text{rank}_{C(\mathbb{T})}(\mathcal{K})$ for every pure Šilov module over $A(\mathbb{D})$ with $C(\mathbb{T})$-extension $\mathcal{K}$. □

For the disk algebra, $\text{rank}_{A(\mathbb{D})}(\mathcal{S}_1) \leq \text{rank}_{A(\mathbb{D})}(\mathcal{S}_2)$ for pure Šilov modules $\mathcal{S}_1 \subseteq \mathcal{S}_2$, as the following theorem shows. By contrast, there are submodules of $H^2(\mathbb{D}^n)$ of arbitrarily high rank over $A(\mathbb{D}^n)$. The following theorem was proved in the rank one case by Beurling [26] and extended to the higher rank cases by Lax [77] and Halmos [68].

<u>Theorem 2.16 (Beurling-Lax-Halmos).</u> Let $\mathcal{M}$ be a submodule of $H^2_{\mathcal{E}}(\mathbb{D})$ with $\text{rank}_{A(\mathbb{D})}(\mathcal{M}) = \dim(\mathcal{E}')$. Then there is a measurable isometry $V(e^{it})$: $\mathcal{E}' \to \mathcal{E}$ defined a.e., such that $\mathcal{M} = V\, H^2_{\mathcal{E}'}(\mathbb{D})$.

<u>Proof.</u> By Theorem 2.11, $\mathcal{M}$ is $A(\mathbb{D})$-module isomorphic to $H^2_{\mathcal{E}'}(\mathbb{D})$ with $\text{rank}_{A(\mathbb{D})}(\mathcal{M})=\dim(\mathcal{E}')$. By Proposition 2.12, this map extends to a $C(\mathbb{T})$-module isometry $V\colon L^2_{\mathcal{E}'}(\mathbb{T}) \to L^2_{\mathcal{E}}(\mathbb{T})$ which by Theorem 1.18 is decomposable. But any decomposable isometry is clearly of the above form. □

Corollary 2.17. Let $\mathcal{S}_1 \subseteq \mathcal{S}_2$ be pure Šilov modules for $A(\mathbb{D})$. Then $\text{rank}_{A(\mathbb{D})}(\mathcal{S}_1) \leq \text{rank}_{A(\mathbb{D})}(\mathcal{S}_2)$.

Proof. Without loss of generality, $\mathcal{S}_2 = H^2{}_{\mathcal{E}}(\mathbb{D})$ and $\mathcal{S}_1 = \mathcal{M}$. Since $V(e^{it})$: $\mathcal{E}' \to \mathcal{E}$ is an isometry, $\text{rank}_{A(\mathbb{D})}(\mathcal{S}_1) = \dim(\mathcal{E}') \leq \dim(\mathcal{E}) = \text{rank}_{A(\mathbb{D})}(\mathcal{S}_2)$. □

The above map $V(e^{it})$ can be shown to be the radial limit almost everywhere of an analytic operator-valued function on $\mathbb{D}$[61]. In the case where $\dim(\mathcal{E}) = 1$, and hence $\dim(\mathcal{E}') = 1$, $\varphi(e^{it}) = V(e^{it})\cdot 1$ is the radial limit of a function in $H^\infty(\mathbb{D})$ with $|\varphi(e^{it})| = 1$ almost everywhere. Such a function is called inner.

One of the key ingredients in the proof of Theorem 2.16 was Proposition 2.13, which said that isometric A-module embeddings of Šilov modules, lift to isometric $C(\partial A)$-module embeddings of their minimal $C(\partial A)$-extensions. For a quite large class of function algebras, every bounded A-module map between Šilov modules lifts to a unique $C(\partial A)$-module map between their respective $C(\partial A)$-extensions. This allows us to identify the bounded A-module maps for these Šilov modules.

Definition 2.18. Let A be a function algebra on X. Then A is approximating in modulus if for every $\varepsilon > 0$ and every positive function g in C(X), there exists f in A such that $\|g - |f|^2\| < \varepsilon$. The algebra A is convexly approximating in modulus if for every $\varepsilon > 0$ and every positive function g in C(X), there exist a finite set $f_1,\ldots,f_n$ in A such that $\| g-(|f_1|^2 + \ldots + |f_n|^2)\| < \varepsilon$.

It is not hard to see that Dirichlet algebras and logmodular algebras are approximating in modulus. Thus, for example $A(\mathbb{D})$ and $H^\infty(\mathbb{D})$ are approximating in modulus on their Šilov boundaries (see [63]).

However, not every function algebra is even convexly approximating in modulus. If we let

$X = \{(z,t) \in \mathbb{C}^2 |\ |z| \leq 1, -1 \leq t \leq +1\}$ and let

$A = \{f \in C(X) |\ f(z,0)$ is analytic for $|z| < 1\}$, then X is

the Šilov boundary of A [99, Example 7.8]. But every function u which is a convex combination of $|f^2|$ for f in A, has the property that $(u_{xx} + u_{yy})(x + iy, 0)$ is positive for $x^2 + y^2 < 1$, and so A can not be convexly approximating in modulus.

The following result of Glicksberg [66] is useful for determining if a function algebra is approximating in modulus.

Proposition 2.19 (Glicksberg). Let A be a function algebra on X and let $U = \{u \in A: |u(x)| = 1$ for all $x \in X\}$. If U separates points on X, then A is approximating in modulus.

Proof. The set $\overline{U}\,A = \{\overline{u}f |\ u \in U, f \in A\}$ is an algebra and consequently contains the algebra generated by U and $\overline{U}$. This latter algebra is uniformly dense in C(X) by the Stone-Weierstrass theorem.

Thus, if g is any positive function on X, then $\sqrt{g}$ can be approximated by $\overline{u}f$ for some f in A and u in U. Hence g is approximated by $|f|^2$. □

The algebras $A(\mathbb{D}^n)$ on $\mathbb{T}^n$ can be easily seen to satisfy Glicksberg's condition. If $\Omega \subseteq \mathbb{C}$ is a bounded open set whose boundary consists of a finite number of disjoint analytic curves, then by [62, Theorem 4.6.5], $R(\Omega-)$ on $\partial\Omega$ also can be seen to satisfy Glicksberg's condition. Thus, all these algebras are approximating in modulus.

The relevance of these algebras is contained in the following commutant lifting theorem of Mlak [83].

Theorem 2.20 (Mlak). Let A be a function algebra that is convexly approximating in modulus on ∂A and let $\mathcal{S}_i$ be a Šilov module for A with minimal $C(\partial A)$-extension $\mathcal{H}_i$, $i = 1, 2$. Then every bounded A-module map $T: \mathcal{S}_1 \to \mathcal{S}_2$ lifts to a bounded $C(\partial A)$-module map $\tilde{T}: \mathcal{H}_1 \to \mathcal{H}_2$ with $\|T\| = \|\tilde{T}\|$.

Proof. Let $\mathcal{H}_i = \int \oplus \mathcal{H}_i(x) d\nu_i(x)$ be the direct integral representation, with Γ_i, their respective fields for $i = 1, 2$. Choose a total set $\{h_n(x)\}$ for $\mathcal{S}_1$ and let $h'_n(x) = Th_n$.

Since $C(\partial A)\cdot\mathcal{S}_1$ is dense in $\mathcal{H}_1$, we have that $\{h_n(x)\}$ is total in $\mathcal{H}_1(x)$ ν_1 almost everywhere, and so h_n is a fundamental sequence for Γ_1. Let $\{k_n(x)\}$ be an enumeration of the finite linear combinations of the set $\{h_n(x)\}$ with rational complex coefficients. Let $\{k'_n(x)\}$ be the corresponding combinations of the $h'_n(x)$'s.

For any f in A, we have that $T(f\cdot k_n) = f(x)k'_n(x)$ ν_2- almost everywhere and hence,

$$\int_X |f(x)|^2 \|k'_n(x)\|^2 d\nu_2(x) \le \|T\|^2 \int_X |f(x)|^2 \|k_n(x)\|^2 d\nu(x)$$

and consequently,

$$\int_X |g(x)| \|k'_n(x)\|^2 d\nu_2(x) \le \|T\|^2 \cdot \int_X |g(x)| \|k_n(x)\|^2 d\nu(x)$$

for every positive g in $C(\partial A)$.

Thus, we see that there exists a (Γ_1,Γ_2)-measurable field of operators $T(x)$: $\mathcal{H}_1(x)\to\mathcal{H}_2(x)$ such that $T(x)\cdot k_n(x)=k'_n(x)$. The field of operators $\{T(x)\}$ defines a bounded operator $\tilde{T}$: $\mathcal{H}_1\to\mathcal{H}_2$, which clearly is a $C(\partial A)$-module map. □

There is only one class of function algebras which includes the disk algebra for which the pure Šilov modules have been completely analyzed. Let Ω be a bounded, finitely connected open subset of $\mathbb{C}$ with $\partial\Omega$ consisting of finitely many disjoint simple, closed, rectifiable curves. It is well-known that the Šilov boundary of $R(\Omega^-)$ can be identified with $\partial\Omega$. The pure Šilov modules for $R(\Omega^-)$ were described by Abrahamse and the first author in [5].

Let μ denote arc-length measure on $\partial\Omega$, and let $H^2(\Omega)$ be the closure of $R(\Omega^-)$ in $L^2(\partial\Omega, \mu)$. If $\mathcal{E}$ is a Hilbert space, then the Hilbert spaces $L_{\mathcal{E}}{}^2(\partial\Omega, \mu)$ and $H^2{}_{\mathcal{E}}(\Omega)$ are defined in a manner analogous to the case of the unit disk, and these latter spaces are pure Šilov modules for $R(\Omega^-)$ with the usual pointwise multiplication. However, if Ω is not simply-connected, these are not all the pure Šilov modules.

Let E be a flat Hermitian holomorphic complex vector bundle over Ω [5], with rank E equalling the dimension of the fiber, possibly infinite. Then E can be extended to be a flat Hermitian holomorphic complex vector bundle over some proper neighborhood of Ω, which we still denote by E. The closure of the holomorphic sections of E in $L^2{}_{E|\partial\Omega}(\partial\Omega, \mu)$ is a contractive Hilbert module over $R(\Omega^-)$, under pointwise multiplication, denoted by $H^2{}_E(\Omega)$. These are pure Šilov modules for $R(\Omega^-)$. The following is proven in [5].

Theorem 2.21 (Abrahamse-Douglas). If $\mathcal{S}$ is a pure Šilov module for $R(\Omega^-)$, then $\mathcal{S}$ is isomorphic to $H^2{}_E(\Omega)$ for some flat Hermitian holomorphic complex vector bundle E over Ω. Moreover, if E_1 and E_2 are two such bundles, then $H^2{}_{E_1}(\Omega)$ and $H^2{}_{E_2}(\Omega)$ are isomorphic if and only if E_1 and E_2 are equivalent as flat Hermitian holomorphic vector bundles and are similar if and only if rank (E_1) = rank (E_2).

Thus, we see that while unitary equivalence detects the finer bundle structure, similarity only detects the module rank. It is not difficult to show that $\mathrm{rank}_{R(\Omega^-)}(H^2{}_E(\Omega))$ = rank(E), which for $A(\mathbb{D})$ was a complete unitary invariant for pure Šilov modules.

The flat unitary holomorphic vector bundles over Ω of rank n, n = 1,2,...,$\aleph_0$ are classified by unitary equivalence classes of unitary representations of the fundamental group of Ω on a Hilbert space of dimension n. Thus, if Ω has k holes, then a bundle is determined by a choice of k unitaries and two such choices will give equivalent bundles if and only if the two sets of unitaries are simultaneously unitarily equivalent [18].

This description of the bundles yields an alternate description of the spaces $H^2{}_E(\Omega)$. Choose k cut lines $C_1, \ldots, C_k$ such that $\Omega \backslash C$, $C = C_1 \cup \ldots \cup C_k$ is simply-connected. Suppose that $\mathcal{E}$ is a Hilbert space with rank(E) = dim($\mathcal{E}$), and that E is determined by unitaries $U_1, \cdots, U_k$ on $\mathcal{E}$. Consider the space of holomorphic $\mathcal{E}$-valued functions on $\Omega \backslash C$ with "jump" discontinuities at each cut line C_k whose values change by a factor of U_k as the variable crosses the cut line in one fixed direction This is a $R(\Omega)$-module under pointwise multiplication and $H^2{}_E(\Omega)$ is module isomorphic to the set of such functions whose limits on $\partial\Omega$ exist a.e. and are square-integrable (see [18]).

When Ω is an annulus and rank(E) = 1, these are precisely the "modulus automorphic" functions studied by Sarason [96].

It is a good problem to try to classify the pure Šilov modules for any natural function algebra. However, at present, most would seem to be beyond our reach. For example, the Šilov modules for $A(\mathbb{D}^2)$ are connected with the representing measures for points, a very complex problem.

We have already seen one example of a Šilov module for $A(\mathbb{D}^2)$, namely the Hardy module $H^2(\mathbb{D}^n)$. To illustrate some other examples, consider the following circles in $\mathbb{T}^2$:

$$C_1 = \{(z_1,z_2): \ z_1 = 1, \ \ |z_2| = 1\},$$

$$C_2 = \{(z_1,z_2): \ z_2 = 1, \ \ |z_1| = 1\},$$

$$C_3 = \{(z_1,z_2): \ z_1 = z_2, \ |z_1| = 1\}.$$

If we let μ_i denote normalized arc-length measure on C_i, then the closure of $A(\mathbb{D}^2)$ in $L^2(\mu_i)$ yields a pure Šilov module for $A(\mathbb{D}^2)$, $i = 1,2,3$. Moreover, all of these modules are distinct and of rank 1.

A more limited problem is the attempt to classify the Hilbert submodules of $H^2(\mathbb{D}^n)$ for $A(\mathbb{D}^n)$. When $n=1$, we see that all these submodules are equivalent via the Beurling-Lax-Halmos theorem for $H^2(\mathbb{D})$. This is not the case for $n > 1$. We illustrate this using a class of Hilbert submodules of $H^2(\mathbb{D}^n)$ studied by Ahern-Clark [10].

If I is an ideal in $\mathbb{C}[z_1,\ldots,z_n]$, then the norm closure of I in $H^2(\mathbb{D}^n)$ is a Hilbert submodule of $H^2(\mathbb{D}^n)$ over $A(\mathbb{D}^n)$. One can show that not all submodules are of this form, since modules of this form have finite rank because $\mathbb{C}[z_1,\ldots,z_n]$ is Noetherian and examples of submodules of $H^2(\mathbb{D}^n)$ of infinite rank are known [94]. For I an ideal in $\mathbb{C}[z_1,\ldots,z_n]$, let Z(I) denote the common zero set of the polynomials in I. Ahern and Clark characterize the submodules of $H^2(\mathbb{D}^n)$ which are the closures of ideals having a finite zero set contained in $\mathbb{D}^n$.

Before describing their result, we introduce the following important concept.

Definition 2.22. If $\mathcal{H}$ is a Hilbert module over A and $\mathcal{M}$ is a submodule, then we form the quotient module, $\mathcal{H}/\mathcal{M}$, which is a Hilbert module over A. The quotient space $\mathcal{H}/\mathcal{M}$ is a Hilbert space since of course it can be identified with $\mathcal{M}^{\perp}$. The module action is given by $f\cdot(h+\mathcal{M}) = f\cdot h + \mathcal{M}$.

This action is easily seen to be bounded and, in fact, $K_{\mathcal{H}/\mathcal{M}} \leq K_{\mathcal{H}}$.

Although all submodules of $H^2(\mathbb{D})$ are isomorphic, that is not the case for the quotient modules. It is not hard to see that if $H^2_z(\mathbb{D})$ denotes the submodule of $H^2(\mathbb{D})$ of functions which vanish at the point z for $|z| < 1$, then $H^2(\mathbb{D})/H^2_z(\mathbb{D})$ is isomorphic to $\mathbb{C}_z$.

Theorem 2.23 (Ahern-Clark). If $\mathcal{M}$ is a submodule of $H^2(\mathbb{D}^n)$ of finite codimension, then there is an ideal I in the algebra of polynomials in $\mathbb{C}[z_1,\ldots z_n]$ of finite codimension, such that $\mathcal{M}$ is the closure of I in $L^2(\mathbb{T}^n)$. Moreover, the ideal I can be chosen so that $Z(I)$ is contained in $\mathbb{D}^n$, in which case I is unique. The zero sets of these ideals are a finite subset of $\mathbb{D}^n$, and conversely every ideal I in $\mathbb{C}[z_1,\ldots z_n]$ whose zero set is a finite subset of $\mathbb{D}^n$ has closure in $L^2(\mathbb{T}^n)$ which is a finite codimension submodule of $H^2(\mathbb{D}^n)$.

Proof. Since $\mathbb{C}[z_1,\ldots z_n]$ is dense in $H^2(\mathbb{D}^n)$ and $H^2(\mathbb{D}^n)/\mathcal{M}$ is finite dimensional, the map $\pi : \mathbb{C}[z_1,\ldots,z_n] \to H^2(\mathbb{D}^n)/\mathcal{M}$ is onto and $\ker \pi = \mathcal{M}\cap\mathbb{C}[z_1,\ldots,z_n]$. Setting $I = \ker \pi$ it is clear that I is an ideal and that $I^- \subseteq \mathcal{M}$. But if we write $\mathbb{C}[z_1,\ldots,z_n] = I + \mathcal{F}$ with $\dim(\mathcal{F}) = \dim(H^2(\mathbb{D}^n)/\mathcal{M})$ then $H^2(\mathbb{D}^n) = \mathbb{C}[z_1\ldots,z_n]^- = I^- + \mathcal{F}$, and so

$\dim (H^2(\mathbb{D}^n)/I^-) \leq \dim (\mathcal{F})$ which implies $I^- = \mathcal{M}$.

Now let λ be in $Z(I)$ and let J be the ideal of polynomials which vanish at λ. Since $I \subseteq J$, $\dim (J/I)$ is finite, and $I^- \cap \mathbb{C}[z_1,\ldots,z_n] = I$, it follows that $J^- \cap \mathbb{C}[z_1,\ldots z_n] = J$. However, if λ is not contained in the disk, it is easily shown that $J^- = H^2 (\mathbb{D}^n)$. Hence, it must be the case that λ is in $\mathbb{D}^n$. □

<u>Problem</u> <u>2.23</u>. Analyze the rank 1 submodules of $H^2(\mathbb{D}^n)$. In particular, when is a rank 1 submodule the closure of an ideal in the algebra of polynomials in n-variables?

It is not obvious whether the submodules of $H^2(\mathbb{D}^n)$ determined by different ideals are non-isomorphic. This was first shown to be the case for some ideals by Berger, Coburn, and Lebow [23]. Finally, this question was resolved by the following result of Agrawal-Clark-Douglas [8] which shows just how many distinct submodules of $H^2(\mathbb{D}^n)$ there are. We will not discuss the proof here, since we will prove a more general result in Chapter VI.

<u>Theorem</u> <u>2.24</u> <u>(Agrawal-Clark-Douglas)</u>. If I_1 and I_2 are ideals in $\mathbb{C}[z_1,\ldots,z_n]$ with finite zero sets contained in $\mathbb{D}^n$ with $\mathcal{I}_1$ and $\mathcal{I}_2$ their closures in $L^2(\mathbb{T}^n)$, then these $A(\mathbb{D}^n)$- modules are unitarily equivalent if and only if $I_1 = I_2$.

Thus, submodules of $H^2(\mathbb{D}^n)$ of finite codimension are isomorphic if and only if they are equal. This is not true in general, since for any submodule $\mathcal{I}$ of $H^2(\mathbb{D}^n)$, $\mathcal{I}$ and $z_i\mathcal{I}$ are isomorphic, $i = 1,\ldots,n$.

3 Šilov resolutions and dilations

In the last chapter, we began discussing contractive Hilbert modules over the disk algebra, $A(\mathbb{D})$. We introduced the class of Šilov modules and saw that the theorem of von Neumann-Wold provides models for all Šilov modules for $A(\mathbb{D})$ and determines when they are unitarily equivalent. In this chapter, we shall begin to see how an arbitrary contractive module for $A(\mathbb{D})$ can be described using these Šilov modules.

We adopt the two-step program from homological algebra and algebraic geometry for studying modules. In the first step, a "nice" tractable class of modules is introduced and investigated. In step two, a general module is studied by considering its resolutions in terms of the nice modules. We shall illustrate the possibilities of this approach in the study of contractive Hilbert modules over $A(\mathbb{D})$, using Šilov modules for the nice class. For other function algebras, even when the Šilov modules are fully understood, we are still very far from a description of the contractive Hilbert modules in terms of Šilov modules, or even from understanding when such resolutions exist. This study can be viewed as a reformulation of "dilation" theory and we introduce many of the results from that theory in this chapter.

Recall that an operator $\Phi: \mathcal{H}_0 \to \mathcal{H}_1$ is a <u>partial isometry</u> if Φ acts isometrically from $\mathcal{H}_0 \ominus \ker \Phi$ onto $\operatorname{ran} \Phi = \mathcal{H}_1 \ominus \ker \Phi^*$. A pair of maps $\mathcal{H}_2 \xrightarrow{\Phi_1} \mathcal{H}_1 \xrightarrow{\Phi_0} \mathcal{H}_0$ is <u>exact</u> if $\operatorname{ran} \Phi_1 = \ker \Phi_2$. A sequence of maps is exact if each adjacent pair of maps in the sequence is exact.

A contractive Hilbert module $\mathcal{M}$ for A has a <u>(projective) Šilov resolution</u> if there exist Šilov modules $\mathcal{S}_0, \ldots, \mathcal{S}_n$ for A and partially isometric module maps $\Phi_0,\ldots,\Phi_n$, such that the sequence,

$$(*) \qquad 0 \leftarrow \mathcal{M} \overset{\Phi_0}{\leftarrow} \mathcal{S}_0 \leftarrow \ldots \overset{\Phi_n}{\leftarrow} \mathcal{S}_n \leftarrow 0,$$

is exact.

Since submodules of Šilov modules are Šilov modules, we see that if we let $\mathcal{S}_1' = \Phi_1(\mathcal{S}_1)$, then the Šilov resolution (*) gives rise to a shorter Šilov resolution,

$$(**) \qquad 0 \leftarrow \mathcal{M} \overset{\Phi_0}{\leftarrow} \mathcal{S}_0 \overset{\Phi_1'}{\leftarrow} \mathcal{S}'_1 \leftarrow 0,$$

where Φ'_1 is just the inclusion map. In fact, if we are given an exact sequence, $0 \leftarrow \mathcal{M} \overset{\Phi_0}{\leftarrow} \mathcal{S}_0$, then we obtain (**) by setting $\mathcal{S}'_1 = \ker \Phi_0$.

<u>Definition</u> <u>3.1</u>. A contractive Hilbert module $\mathcal{M}$ for the function algebra A has a <u>Šilov</u> <u>resolution</u> if there exist Šilov modules $\mathcal{S}_0$ and $\mathcal{S}_1$ for A and partially isometric module maps Φ: $\mathcal{S}_0 \to \mathcal{M}$ and Φ_1: $\mathcal{S}_1 \to \mathcal{S}_0$ such that the sequence

$$0 \leftarrow \mathcal{M} \overset{\Phi_0}{\leftarrow} \mathcal{S}_0 \overset{\Phi_1}{\leftarrow} \mathcal{S}_1 \leftarrow 0,$$

is exact. The module $\mathcal{S}_0$ is called a <u>Šilov</u> <u>dominant</u> for $\mathcal{M}$.

We say that the resolution is <u>strongly</u> <u>minimal</u> if there is no submodule $\mathcal{S}'_0$ of $\mathcal{S}_0$ such that the restriction of Φ_0 to $\mathcal{S}'_0$ is a partial isometry and $\Phi_0(\mathcal{S}'_0) = \mathcal{M}$. We say the resolution is <u>weakly</u> <u>minimal</u>, if $\mathcal{S}_0$ can not be split into a direct sum of Šilov modules $\mathcal{S}_0 = \mathcal{S}'_0 \oplus \mathcal{S}_0''$, with the module action on $\mathcal{S}_0$ equal to the direct sum of the two module actions, such that the restriction of Φ_0 to $\mathcal{S}'_0$ is a partial isometry and $\Phi_0(\mathcal{S}'_0) = \mathcal{M}$.

Given any Šilov resolution for $\mathcal{M}$,

$$0 \leftarrow \mathcal{M} \overset{\Phi_0}{\leftarrow} \mathcal{S}_0 \overset{\Phi_1}{\leftarrow} \mathcal{S}_1 \leftarrow 0,$$

if we let $\mathcal{S}'_0$ be the intersection of all submodules $\mathcal{S}$ of $\mathcal{S}_0$ which have the properties that Φ_0 is a partial isometry when restricted to $\mathcal{S}$ and $\Phi_0(\mathcal{S}) = \mathcal{M}$, and if we set $\mathcal{S}'_1 = \Phi_1^{-1}(\mathcal{S}'_0)$, then

$$0 \leftarrow \mathcal{M} \overset{\Phi_0'}{\leftarrow} \mathcal{S}'_0 \overset{\Phi_1'}{\leftarrow} \mathcal{S}'_1 \leftarrow 0$$

is a strongly minimal Šilov resolution for $\mathcal{M}$, where Φ'_i denotes the restriction of Φ'_i to $\mathcal{S}_i$, $i = 1, 2$. Thus, strongly minimal Šilov resolutions exist. We call this resolution the <u>strongly minimal part</u> of the original resolution.

Clearly, strongly minimal Šilov resolutions are weakly minimal, and so these always exist as well.

We call Šilov dominants for $\mathcal{M}$ strongly (respectively, weakly) minimal if the Šilov resolution they give rise to is strongly (respectively, weakly) minimal.

In terms of diagrams, to say that the resolution is strongly minimal means that if we are given a commuting diagram of the form,

$$\begin{array}{ccccccccc}
 & & 0 & & & & & & \\
 & & \uparrow & & & & & & \\
0 & \leftarrow & \mathcal{M} & \leftarrow & \mathcal{S}_0 & \leftarrow & \mathcal{S}_1 & \leftarrow & 0 \\
 & & \uparrow & & \uparrow & & \uparrow & & \\
0 & \leftarrow & \mathcal{M} & \leftarrow & \mathcal{S}'_0 & \leftarrow & \mathcal{S}'_1 & \leftarrow & 0 \\
 & & \uparrow & & \uparrow & & \uparrow & & \\
 & & 0 & & 0 & & 0 & &
\end{array}$$

where the arrows represent partially isometric module maps, and all sequences are exact, then the maps $\mathcal{S}'_i \to \mathcal{S}_i$ are necessarily onto, $i = 0,1$.

Both types of resolutions will play a role in our considerations. We begin with an example of a weakly minimal Šilov resolution which is not strongly minimal.

We adopt the convention, in this chapter, that unlabeled arrows always indicate partially isometric module maps.

<u>Example 3.2</u>. Let Ω be an analytic Cauchy domain in $\mathbb{C}$, so that $\partial\Omega$ consists of n + 1 disjoint analytic, simple, closed curves. Let ds denote arc length measure on $\partial\Omega$ and let $H^2(\partial\Omega,ds)$ be the closure in $L^2(\partial\Omega,ds)$ of the algebra $R(\Omega^-)$ of bounded, rational functions on Ω. This is a Šilov module for $R(\Omega^-)$.

If we fix a point ω in Ω, then the point evaluation $\Phi(g) = g(\omega)$ is a bounded, linear functional on $H^2(\partial\Omega,ds)$, and hence a bounded $R(\Omega^-)$-module map onto $\mathbb{C}_\omega$. The kernel of this map is $H^2_\omega(\partial\Omega,ds) = \{g \in H^2(\partial\Omega,ds): g(\omega) = 0\}$.

The exact sequence of contractive, Hilbert $R(\Omega^-)$- modules,

$$0 \leftarrow \mathbb{C}_\omega \overset{\Phi_0}{\leftarrow} H^2(\partial\Omega,ds) \leftarrow H^2_\omega(\partial\Omega,ds) \leftarrow 0$$

with $\Phi_0 = \Phi/\|\Phi\|$, is easily seen to be a Šilov resolution. Since $H^2(\partial\Omega,ds)$ can not be decomposed into a module direct sum of $R(\Omega^-)$- modules, that is, with the direct sum of the $R(\Omega^-)$-actions, this resolution is weakly minimal.

Let k_ω be the vector in $H^2(\partial\Omega,ds)$ such that $g(\omega)=\langle g, k_\omega\rangle$. This is called the <u>Szego kernel</u> for ω, and it is known that this function vanishes at n points in Ω [25]. Call these $\lambda_1,\ldots\lambda_n$, and set

$$H^2_\lambda(\partial\Omega,ds) = \{g \in H^2(\partial\Omega,ds): g(\lambda_i) = 0, i = 1, \ldots, n\},$$

$$H^2_{\lambda,w}(\partial\Omega,ds) = \{g \in H^2_\lambda(\partial\Omega,ds) : g(\omega) = 0\}.$$

Since k_λ is in $H^2_\lambda(\partial\Omega, ds)$, Φ_0 is a partial isometry on this space and maps it onto $\mathbb{C}_\omega$. Thus, our original Šilov resolution is not strongly minimal.

However, the Šilov resolution

$$0 \leftarrow \mathbb{C}_\omega \overset{\Phi_0}{\leftarrow} H^2_\lambda(\partial\Omega, ds) \leftarrow H^{\;2}_{\lambda,\omega}(\partial\Omega, ds) \leftarrow 0,$$

is strongly minimal. To see this, note that the partial isometry onto $\mathbb{C}_\omega$ is given by taking the inner product with $k_\omega/\|k_\omega\|$. Thus, for this map to restrict to a partial isometry on a submodule of $H^2_\lambda(\partial\Omega, ds)$, it is necessary that the submodule contain k_ω. But by standard theorems on the factorization of analytic functions, the subspace $R(\Omega^-)\cdot k_\omega$ is dense in $H^2_\lambda(\partial\Omega, ds)$, and so there is no proper submodule of $H^2_\lambda(\partial\Omega, ds)$ containing k_ω.

In order to better understand strongly minimal resolutions, we need to first make some comments on quotient modules. Recall that if $\mathcal{S}_0$ is any Hilbert module for A, with submodule $\mathcal{S}_1$, then we may form a quotient module $\mathcal{S}_0/\mathcal{S}_1 = \mathcal{M}$ with action $f\cdot(h + \mathcal{S}_1) = f\cdot h + \mathcal{S}_1$. Clearly, as Hilbert spaces, we can regard $\mathcal{M}$ as a subspace of $\mathcal{S}_0$; in fact, $\mathcal{S}_1 \oplus \mathcal{M} = \mathcal{S}_0$. With respect to this identification, the module action on $\mathcal{M}$ is given by $(f,h) \to P_{\mathcal{M}}(f\cdot h)$, where $f\cdot h$ is the module action on $\mathcal{S}_0$ and $P_{\mathcal{M}}$ is the orthogonal projection of $\mathcal{S}_0$ onto $\mathcal{M}$. We call this action on $\mathcal{M}$ the <u>compression</u> of the action on $\mathcal{S}_0$ to the subspace $\mathcal{M}$.

<u>Theorem</u> <u>3.3</u>. Let $\mathcal{M}$ be a contractive Hilbert module for the function algebra A, and let $0 \leftarrow \mathcal{M} \overset{\Phi_0}{\leftarrow} \mathcal{S}_0 \overset{\Phi_1}{\leftarrow} \mathcal{S}_1 \leftarrow 0$ be a Šilov resolution for $\mathcal{M}$, and regard $\mathcal{M}$ as a subspace of $\mathcal{S}_0$. If we let $\mathcal{S}'_0$ be the cyclic Hilbert space generated by $\mathcal{M}$, that is, the closed span of $\{f\cdot h \mid f \in A,\ h \in \mathcal{M}\}$, under the module action on $\mathcal{S}_0$, and $\mathcal{S}'_1 = \mathcal{S}'_0 \ominus \mathcal{M}$, then $0 \leftarrow \mathcal{M} \leftarrow \mathcal{S}'_0 \leftarrow \mathcal{S}'_1 \leftarrow 0$ is a

strongly minimal Šilov resolution for $\mathcal{M}$. Moreover, the original resolution is strongly minimal if and only if $\mathcal{S}'_0 = \mathcal{S}_0$.

Proof. With our identifications, the partial isometry from $\mathcal{S}_0$ to $\mathcal{M}$ is just $P_{\mathcal{M}}$. This restricts to a partial isometry on $\mathcal{S}'_0$ since $\mathcal{S}'_0$ contains $\mathcal{M}$.

Thus, $0 \leftarrow \mathcal{M} \leftarrow \mathcal{S}'_0 \leftarrow \mathcal{S}'_1 \leftarrow 0$ is a Šilov resolution, and so the original Šilov resolution is strongly minimal if and only if $\mathcal{S}'_0 = \mathcal{S}_0$. This new resolution is strongly minimal since any submodule of $\mathcal{S}'_0$ for which $P_{\mathcal{M}}$ is a partial isometry onto $\mathcal{M}$ contains the cyclic module generated by $\mathcal{M}$. $\square$

This quotient module approach also is useful for identifying an important case where weakly minimal Šilov resolutions are strongly minimal.

Recall that a function algebra A is <u>Dirichlet</u> if $A + \overline{A}$ is dense in $C(\partial A)$.

Proposition 3.4. Let A be a Dirichlet algebra, and let $\mathcal{M}$ be a contractive Hilbert module for A. Then any weakly minimal Šilov resolution for A is strongly minimal.

Proof. Let $0 \leftarrow \mathcal{M} \leftarrow \mathcal{S}_0 \leftarrow \mathcal{S}_1 \leftarrow 0$ be a weakly minimal Šilov resolution, and regard $\mathcal{M}$ as a subspace of $\mathcal{S}_0$ with the compressed action. We must determine if
$\mathcal{S}'_0 = A \cdot \mathcal{M} \equiv \{\sum_i f_i \cdot h_i | f_i \in A, h_i \in \mathcal{M}\}^-$ is equal to $\mathcal{S}_0$, where the module product is taken in $\mathcal{S}_0$. Let $\mathcal{H}$ be the minimal $C(\partial A)$-extension of $\mathcal{S}_0$, so that for each g in $C(\partial A)$, the map defined by $M_g h = g \cdot h$ on $\mathcal{H}$ is a normal operator.

Assume $\mathcal{S}'_0 \neq \mathcal{S}_0$ and let h in $\mathcal{S}_0$ be orthogonal to $\mathcal{S}'_0$. Then for f in A and k in $\mathcal{M}$, we have

$$\langle h, M_{\bar{f}}\, k\rangle_{\mathcal{H}} = \langle h, M_f^*\, k\rangle_{\mathcal{H}} = \langle M_f h, k\rangle_{\mathcal{H}} = 0$$

since h is in $\mathcal{S}_1$ which is a submodule for A. Also, since $\mathcal{S}'_0$ is a submodule for A containing $\mathcal{M}$, $\langle h, M_f\, k\rangle = 0$ for k in $\mathcal{M}$.

Hence, $C(\partial A)\cdot\mathcal{M} = \mathcal{H}'$ is a subspace of $\mathcal{H}$ orthogonal to $\mathcal{H}$ and contains $\mathcal{S}'_0$. Since h was arbitrary, we have that $\mathcal{S}_0'' = \mathcal{S}_0 \ominus \mathcal{S}'_0$ is contained in $\mathcal{H} \ominus \mathcal{H}' = \mathcal{H}''$, while $\mathcal{S}'_0$ is contained in $\mathcal{H}'$.

Finally, since $\mathcal{H}'$ is $C(\partial A)$-invariant, we have that $\mathcal{H} \ominus \mathcal{H}' = \mathcal{H}''$ defines a splitting of $\mathcal{H}$ into a direct sum of $C(\partial A)$-modules, with the direct sum action. Thus, $\mathcal{S}_0 = \mathcal{S}'_0 \oplus \mathcal{S}_0''$ is a splitting of $\mathcal{S}_0$, with the direct sum action. However, since $\mathcal{M}$ is contained in $\mathcal{S}'_0$, this contradicts the weak minimality, and so $\mathcal{S}_0' = \mathcal{S}_0$. □

We now wish to connect the theory of Šilov resolutions to the theory of dilations.

We saw earlier, when studying quotient modules, an instance of a subspace $\mathcal{M}$ of a bounded Hilbert module for A such that the pairing $(f,h) \to P_{\mathcal{M}}(f\cdot h)$ for f in A and h in $\mathcal{M}$ defined a module action on $\mathcal{M}$. For general subspaces this is not the case; associativity of the module product fails. If we define $T_f\colon \mathcal{M} \to \mathcal{M}$ by, $T_f(h) = P_{\mathcal{M}}(f\cdot h)$, for f in A, then, in general $T_{f_1}\cdot T_{f_2} \neq T_{f_1 f_2}$. That is, the compression of a module action need not be a module action. It is a module action if and only if

$$P_{\mathcal{M}}(f_1 f_2\cdot h) = P_{\mathcal{M}}\Big(f_1\cdot\ (P_{\mathcal{M}}(f_2\cdot h))\ \Big)$$

<u>Definition 3.5</u>. Let $\mathcal{H}$ be a bounded Hilbert module for A, and let $\mathcal{M}$ be a subspace of $\mathcal{H}$. If the compression $(f,h) \to P_{\mathcal{M}}(f\cdot h)$ for f in A and h in $\mathcal{M}$ of the module action on $\mathcal{M}$ to $\mathcal{M}$, defines a module action on $\mathcal{M}$, then we say that $\mathcal{M}$ is a <u>semisubmodule</u> of $\mathcal{H}$ for A. If A is contained in C(X) and $\mathcal{H}$

is also a contractive Hilbert C(X)-module, then we say that $\mathcal{H}$ is an X-dilation of $\mathcal{M}$.

Although every submodule of a module is a semisubmodule, the converse is usually not valid. The following theorem of Sarason [97] characterizes semisubmodules and leads to the correspondence between X-dilations and Šilov resolutions.

Theorem 3.6 (Sarason). Let $\mathcal{H}$ be a bounded Hilbert module for A and let $\mathcal{M}$ be a subspace of $\mathcal{H}$. Then $\mathcal{M}$ is a semisubmodule of $\mathcal{H}$ for A if and only if there is a nested pair $\mathcal{S}_1 \subseteq \mathcal{S}_0$ of submodules of $\mathcal{H}$ such that $\mathcal{S}_1 \oplus \mathcal{M} = \mathcal{S}_0$ as spaces.

Proof. Assuming the latter condition, we know that we can identify $\mathcal{M}$ with the quotient module $\mathcal{S}_0/\mathcal{S}_1$. With these identifications, the module action on $\mathcal{M}$ is given by $(f,h) \to P^0_{\mathcal{M}}(f \cdot h)$ where $P^0_{\mathcal{M}}$ is the projection of $\mathcal{S}_0$ onto $\mathcal{M}$, and $f \cdot h$ is the module action on $\mathcal{S}_0$. However, since $\mathcal{S}_0$ is a submodule of $\mathcal{H}$, $P^0_{\mathcal{M}}(f \cdot h) = P_{\mathcal{M}}(f \cdot h)$ where $P_{\mathcal{M}}$ is the projection of $\mathcal{H}$ onto $\mathcal{M}$, and $f \cdot h$ is the module action on $\mathcal{H}$. Thus, $P_{\mathcal{M}}(f \cdot h)$ defines a module action on $\mathcal{M}$, and so $\mathcal{M}$ is a semisubmodule.

Conversely, assume that $\mathcal{M}$ is a semisubmodule of $\mathcal{H}$ and let $\mathcal{S}_0$ be the cyclic submodule of $\mathcal{H}$ generated by $\mathcal{M}$, that is, $\mathcal{S}_0$ is the closure of $A \cdot \mathcal{M}$. We need only show that $\mathcal{S}_1 = \mathcal{S}_0 \ominus \mathcal{M}$ is also a submodule of $\mathcal{H}$.

Let h in $\mathcal{S}_0$ be of the form $h = \sum_i f_i \cdot k_i$, where f_i is in A and k_i in $\mathcal{M}$, the module action is in $\mathcal{S}_0$ and the sum is finite. Then $h - P_{\mathcal{M}}h$ is in $\mathcal{S}_1$ and for f in A, we have

$$P_{\mathcal{M}}\big(f(h-P_{\mathcal{M}}h)\big) = P_{\mathcal{M}}fh - P_{\mathcal{M}}fP_{\mathcal{M}}h = P_{\mathcal{M}}fh - \sum P_{\mathcal{M}}fP_{\mathcal{M}}f_ik_i =$$

$$P_{\mathcal{M}}fh - \sum P_{\mathcal{M}}ff_ik_i = 0.$$

Thus $f\cdot((h-P_{\mathcal{M}}h)$ is in $\mathcal{S}_0 \ominus \mathcal{M} = \mathcal{S}_1$ and so $\mathcal{S}_1$ is invariant under the module action, that is, $\mathcal{S}_1$ is a submodule. □

<u>Corollary</u> <u>3.7</u>. Let $\mathcal{M}$ be a contractive Hilbert module for A. Then $\mathcal{M}$ has a Šilov resolution if and only if $\mathcal{M}$ has a ∂A-dilation.

<u>Proof</u>. If $\mathcal{M}$ has a Šilov resolution $0 \leftarrow \mathcal{M} \leftarrow \mathcal{S}_0 \leftarrow \mathcal{S}_1 \leftarrow 0$, then we have seen that $\mathcal{M}$ may be regarded as a semisubmodule of $\mathcal{S}_0$ for A. Since $\mathcal{S}_0$ is Šilov, it has a $C(\partial A)$-extension $\mathcal{H}$. Clearly, $\mathcal{M}$ is also a semisubmodule of $\mathcal{H}$ for A and so $\mathcal{H}$ is a ∂A-dilation.

Conversely, if $\mathcal{M}$ has a ∂A-dilation $\mathcal{H}$, then by Sarason's theorem, there are submodules $\mathcal{S}_1$ and $\mathcal{S}_0$ of $\mathcal{H}$ for A, such that $\mathcal{M}$ is the quotient of $\mathcal{S}_0$ by $\mathcal{S}_1$. Thus, $0 \leftarrow \mathcal{M} \leftarrow \mathcal{S}_0 \leftarrow \mathcal{S}_1 \leftarrow 0$ is a Šilov resolution of $\mathcal{M}$. □

The correspondence between Šilov resolutions and ∂A-dilations leads to a third notion of minimality, which is borrowed from dilation theory.

<u>Definition</u> <u>3.8</u>. Let $\mathcal{M}$ be a contractive Hilbert module for A, with Šilov resolution,

$$0 \leftarrow \mathcal{M} \leftarrow \mathcal{S}_0 \leftarrow \mathcal{S}_1 \leftarrow 0,$$

and let $\mathcal{H}$ be the unique minimal $C(\partial A)$-extension of $\mathcal{S}_0$. Regard $\mathcal{M}$ as a semisubmodule of $\mathcal{H}$. We say that the Šilov resolution is <u>minimal</u> if $\mathcal{H}$ is the closure of $C(\partial A) \cdot \mathcal{M}$, that is, if no proper submodule for $C(\partial A)$ of $\mathcal{H}$ contains $\mathcal{M}$.

For the reader familiar with dilation theory, we remark that we have chosen our language such that minimal Šilov resolutions correspond to minimal ∂A-dilations as defined in dilation theory. It is important to observe that the correspondence is not one-to-one. The weakly minimal and strongly minimal Šilov resolutions of Example 3.2 are both minimal, and both give rise to the $C(\partial\Omega)$-dilation, with $\mathcal{H} = L^2(\partial\Omega, ds)$. This happens because the minimal $C(\partial\Omega)$-extensions of $H^2(\partial\Omega, ds)$ and $H^2_\lambda(\partial\Omega, ds)$ are both $L^2(\partial\Omega, ds)$. Given a $C(\partial A)$-dilation for $\mathcal{M}$, the proof of Corollary 3.7 gives a canonical way of obtaining a Šilov resolution of $\mathcal{M}$, which is strongly minimal since $\mathcal{M}$ is cyclic in $\mathcal{S}_0$.

The reader may be wondering, why we bother with anything besides strongly minimal resolutions. The reason is that "natural" examples tend not to be cyclic. As we saw in Example 3.2, to obtain the strongly minimal resolution, we had to consider the "unnatural" space $H^2_\lambda(\partial\Omega, ds)$.

We present another example which illustrates the difficulties of defining strongly minimal resolutions.

Example 3.9 (Abrahamse). Let ω be a point in the annulus $\mathbb{A}$ and let $\mathbb{C}_\omega$ be the 1-dimensional contractive Hilbert module over $R(\mathbb{A})$ given by point evaluation at ω. We shall exhibit a family of inequivalent minimal Šilov resolutions of $\mathbb{C}_\omega$ indexed by the unit circle.

Let $\pi : \mathbb{D} \to \mathbb{A}$ be a covering map of $\mathbb{A}$ by $\mathbb{D}$ such that $\pi^{-1}(\omega) = \{z_n\}_{n=-\infty}^{+\infty}$. It is well-known that $\{z_n\}$ is a Blaschke sequence in $\mathbb{D}$ and we let β denote the Blaschke product of these zeroes. Let γ denote the generating deck transformation for π, that is, γ is a linear fractional map on $\mathbb{D}$ such that the orbits of γ are exactly the inverse images of points in $\mathbb{A}$ under π.

Since $\beta \circ \gamma$ is another Blaschke product with the same zeroes as β, there exists λ, $|\lambda| = 1$, such that $\beta \circ \gamma = \lambda\beta$.

For arbitrary μ, $|\mu| = 1$, consider the subspace of $H^2(\mathbb{D})$ consisting of those functions with the property that $f \circ \gamma = \mu f$ which we denote by $H^2{}_\mu(\mathbb{D},\gamma)$. This space can be identified with the space $H^2{}_{E_\mu}(\mathbb{A})$, where E_μ is a flat Hermitian holomorphic bundle over $\mathbb{A}$ of rank 1 with "twist" μ, introduced in chapter 2, and is a pure Šilov module. The action of a function f in $R(\mathbb{A})$ on g in $H^2{}_\mu(\mathbb{D},\gamma)$ is given by $(f\cdot g)(z)=f(\pi(z))g(z)$.

Multiplication by β carries $H^2{}_\mu(\mathbb{D},\gamma)$ into $H^2{}_{\lambda\mu}(\mathbb{D},\gamma)$ and since $|\beta(z)| = 1$ for $|z|=1$, this map defines the isometry, $\tilde{\beta}:H^2{}_\mu(\mathbb{D},\gamma) \rightarrow H^2{}_{\lambda\mu}(\mathbb{D},\gamma)$, which is easily seen also to be a $R(\mathbb{A})$-module map.

We claim that $0 \leftarrow \mathbb{C}_\omega \overset{\Phi}{\leftarrow} H^2{}_{\lambda\mu}(\mathbb{D}, \gamma) \overset{\tilde{\beta}}{\leftarrow} H^2{}_\mu(\mathbb{D}, \gamma) \leftarrow 0$, is a weakly minimal Šilov resolution, where Φ is just the quotient map. To see this, first note that if g is in $H^2{}_{\lambda\mu}(\mathbb{D},\gamma)$ and $g(z_n) = 0$ for some n, then it is zero for all n. Hence $g = \beta g_1$, and necessarily g_1 is in $H^2{}_{\gamma\mu}(\mathbb{D},\gamma)$. Thus $\tilde{\beta}\ H^2{}_\mu(\mathbb{D},\gamma) = \{ g \in H^2{}_{\lambda\mu}(\mathbb{D},\gamma)\colon g(z_n) = 0 \text{ for all } n\}$ and is of codimension at most 1. To see that $\tilde{\beta}H^2{}_\mu(\mathbb{D},\gamma)$ has codimension 1, it is enough to construct a function k_λ in $H^2{}_{\lambda\mu}(\mathbb{D},\gamma)$ with $k_\lambda(z_o) = 1$, which we leave to the reader.

To determine the quotient action, note that for f in $R(\mathbb{A})$ and g in $H^2{}_{\lambda\mu}(\mathbb{D},\gamma)$, $[(f-f(\omega)1)\cdot g](z_n) = (f(\pi(z_n))-f(\omega))g(z_n) = 0$ and hence $(f-f(\omega)\cdot 1)\cdot g$ is in $\tilde{\beta}\ H^2{}_\mu(\mathbb{D},\gamma)$. Thus, in the quotient $f\cdot[g] = f(\omega)\cdot[g]$, where [g] is the coset that contains g. Thus, we do have a Šilov resolution of $\mathbb{C}_\omega$.

If we had an isomorphism between two of these Šilov resolutions, then in particular we would have an $R(\mathbb{A})$-module isomorphism between $H^2{}_\mu(\mathbb{D},\gamma)$ and $H^2{}_{\mu'}(\mathbb{D},\gamma)$ with $\mu\neq\mu'$, which is easily seen to be impossible. Thus, the resolutions in this family are all non-isomorphic..

To see that these resolutions are weakly minimal, it is sufficient to see that for no μ is there a submodule $\mathcal{M}$ of

$H^2{}_{\lambda\mu}(\mathbb{D},\gamma)$ such that $\mathcal{M}^{\perp}$ is also a submodule. But the spaces $H^2{}_{\lambda\mu}(D,\gamma)$ are all "irreducible", and hence the Šilov resolution must be weakly minimal.

However, to be strongly minimal, we would need that $R(\mathbb{A}) \cdot k_\lambda$ is dense in $H^2{}_{\lambda\mu}(\mathbb{D},\gamma)$, which requires delicate knowledge about k_λ. Surprisingly, whether or not the resolution is strongly minimal depends on the choice of λ. See [92] where it is argued that there is a closed arc on the unit circle such that the resolution is strongly minimal on that arc, and not strongly minimal on the complement of the arc.

These resolutions are all in fact minimal, but to see that requires the introduction of the spaces $L^2(\mathbb{D},\gamma)$, see [62].

We now show that the language we have used for these three concepts of minimality is consistent.

<u>Proposition</u> <u>3.10</u>. Let $\mathcal{M}$ be a contractive Hilbert module for A. Then every strongly minimal Šilov resolution of $\mathcal{M}$ is minimal, and every minimal Šilov resolution of $\mathcal{M}$ is weakly minimal.

<u>Proof</u>. Let $0 \leftarrow \mathcal{M} \leftarrow \mathcal{S}_0 \leftarrow \mathcal{S}_1 \leftarrow 0$ be a stronly minimal Šilov resolution of $\mathcal{M}$, regard $\mathcal{M}$ as a semisubmodule of $\mathcal{S}_0$ and let $\mathcal{H}$ be the minimal $C(\partial A)$-extension of $\mathcal{S}_0$. Recall that this means that the closure of $C(\partial A)\mathcal{S}_0$ is $\mathcal{H}$. By the strong minimality of the resolution, $\mathcal{S}_0$ is the closure of $A\cdot\mathcal{M}$. Therefore, the closure of $C(\partial A)\cdot\mathcal{M}$ contains the closure of $C(\partial A)\cdot\mathcal{S}_0$ which is $\mathcal{H}$, and so the resolution is minimal.

Next assume that the resolution is not weakly minimal. Let $\mathcal{S}_0 = \mathcal{S}_0' \oplus \mathcal{S}_0''$ with a direct sum action, and $\mathcal{M} \subseteq \mathcal{S}_0'$. Then we see that the minimal $C(\partial A)$-extension $\mathcal{H}$ of $\mathcal{S}_0$ splits into $\mathcal{H} = \mathcal{H}' \oplus \mathcal{H}$

with $\mathcal{H}'$ the closure of $C(\partial A)\cdot\mathcal{S}_0'$, $\mathcal{H}''$ the closure of $C(\partial A)\cdot\mathcal{S}_0''$, and $\mathcal{M}$ is contained in $\mathcal{H}'$. Hence, the resolution is not minimal. □

Proposition 3.11. Let A be a Dirichlet algebra. Then the three definitions of minimality coincide.

Proof. Apply propositions 3.9 and 3.4. □

At present, we do not know of any examples of weakly minimal Šilov resolutions which are not minimal, but we believe that such examples exist.

Problem 3.12. Construct a weakly minimal Šilov resolution which is not minimal.

Problem 3.13 Do there exist function algebras for which two of the notions of minimality coincide, but not all three?

Now that we have developed the correspondence between Šilov resolutions and dilations, we find that there is quite a bit of information available about these resolutions. The classification of contractive Hilbert modules over A($\mathbb{D}$) follows from a result of Sz-Nagy [100].

Definition 3.14. Let $\mathcal{M}$ be a contractive Hilbert module over A and let $0 \leftarrow \mathcal{M} \leftarrow \mathcal{S}_i \leftarrow \mathcal{R}_i \leftarrow 0$ be two Šilov resolutions of $\mathcal{M}$, $i = 1,2$. We say that these resolutions are isomorphic if there exist module isomorphisms $\phi: \mathcal{S}_1 \rightarrow \mathcal{S}_2$, $\psi: \mathcal{R}_1 \rightarrow \mathcal{R}_2$ such that the following diagram commutes,

$$\begin{array}{ccccccccc} 0 & \leftarrow & \mathcal{M} & \leftarrow & \mathcal{S}_1 & \leftarrow & \mathcal{R}_1 & \leftarrow & 0 \\ & & \mathrm{id}\downarrow & & \phi\downarrow & & \psi\downarrow & & \\ 0 & \leftarrow & \mathcal{M} & \leftarrow & \mathcal{S}_2 & \leftarrow & \mathcal{R}_2 & \leftarrow & 0, \end{array}$$

where id denotes the identity map.

Theorem 3.15 (Sz.-Nagy). Every contractive Hilbert module $\mathcal{M}$ over $A(\mathbb{D})$ has a Šilov resolution. Moreover, any two minimal Šilov resolutions of $\mathcal{M}$ are isomorphic.

Proof. The existence of the Šilov resolution follows from the existence of a unitary dilation, that is, $\partial\mathbb{D}$-dilation, for every contraction operator [100] and the correspondences between contraction operators and contractive Hilbert modules for $A(\mathbb{D})$ and between Šilov resolutions and dilations. The isomorphism between any two minimal Šilov resolutions follows from the fact that any two unitary dilations of a contraction operator are unitarily equivalent [100]. □

In the previous chapter, we discussed the fact that classifying all contractive Hilbert modules for $A(\mathbb{D})$ is equivalent to classifying all contraction operators up to unitary equivalence. However, we now see that Sz.-Nagy's theorem tells us that each of these modules possesses a unique (up to isomorphism), minimal Šilov resolution and we have seen earlier that the Šilov modules for $A(\mathbb{D})$, especially the pure ones, have a simple classification (the von Neumann-Wold theorem). Thus, we see that in some sense all the information for classifying contraction operators is contained in the two maps Φ_0 and Φ_1 in the Šilov resolution $0 \leftarrow \mathcal{M} \overset{\Phi_0}{\leftarrow} \mathcal{S}_0 \overset{\Phi_1}{\leftarrow} \mathcal{S}_1 \leftarrow 0$. If we adopt the convention that Φ_0 is always just the quotient map, then the embedding Φ_1 contains all the information necessary to determine the module $\mathcal{M}$, that is, its corresponding contraction, up to isomorphism. This point of view is the basis for Sz.-Nagy and Foias' theory of characteristic operator functions [101]. The goal of this theory is to extract information about $\mathcal{M}$ from Φ_1. If we restrict to the case where $\mathcal{S}_0$ and $\mathcal{S}_1$ are pure Šilov modules, then $\mathcal{S}_i$ is canonically isomorphic to $H^2_{\mathcal{E}_i}(\mathbb{D})$, $i = 1,2$, and by Theorem 2.15, Φ_1 is given by multiplication by a function θ where $\theta(e^{it})$: $\mathcal{E}_1 \to \mathcal{E}_2$ is an isometry. This isometry is the radial limit of a function analytic on $\mathbb{D}$, and it is that function which Sz.-Nagy and Foias call the characteristic operator function.

For a general function algebra, it is unreasonable to expect uniqueness of minimal, or even strongly minimal, Šilov resolutions. For example, we produced two minimal Šilov resolutions of the point evaluation module C_ω for $R(\Omega^-)$ in example 3.2. Since one of these resolutions was only weakly minimal while the other was strongly minimal, they are clearly not isomorphic. Example 3.9 gives uncountably many non-isomorphic strongly minimal Šilov resolutions.

We now summarize some of the material on the existence of Šilov resolutions which can be found in the dilation theory literature.

__Theorem 3.16 (Arveson__ [12]). Every contractive Hilbert module over a Dirichlet algebra has a unique, up to isomorphism, minimal Šilov resolution.

__Theorem 3.17 (Agler__ [7]). Every contractive Hilbert module for $R(\mathbb{A})$ has a Šilov resolution.

__Theorem 3.18 (Ando__ [88]). Every contractive Hilbert module for $A(\mathbb{D}^2)$ has a Šilov resolution.

__Theorem 3.19 (Parrott__ [85]). There exist contractive Hilbert modules over $A(\mathbb{D}^n)$, $n \geq 3$, which have no Šilov resolutions.

We have already seen examples of the non-uniqueness of Šilov resolutions for $R(\mathbb{A})$. Šilov resolutions for $A(\mathbb{D}^2)$ are also non-unique as the following example shows.

<u>Example 3.20</u>. Let $\mathbb{C}_{(0,0)}$ be the 1-dimensional contractive Hilbert module for $A(\mathbb{D}^2)$ given by point evaluation at (0,0).

The Šilov resolution given by

$0 \leftarrow \mathbb{C}_{(0,0)} \leftarrow H^2(\mathbb{D}^2) \leftarrow H^2_{(0,0)}(\mathbb{D}^2) \leftarrow 0$, where

$H^2_{(0,0)}(\mathbb{D}^2) = \{g \in H^2(\mathbb{D}) | g(0,0) = 0\}$, is easily seen to be strongly minimal.

For a second Šilov resolution consider $C = \{(z_1, z_2): z_1 = z_2, |z_1| = 1\}$, let μ denote normalized arc-length measure on C, let $H^2(\mu)$ be the closure of $A(\mathbb{D}^2)$ in $L^2(\mu)$ and let $H^2_{(0,0)}(\mu) = \{g \in H^2(\mu): g(0,0) = 0\}$. Then,

$$0 \leftarrow \mathbb{C}_{(0,0)} \leftarrow H^2(\mu) \leftarrow H^2_{(0,0)}(\mu) \leftarrow 0$$

is a second strongly minimal Šilov resolution of $\mathbb{C}_{(0,0)}$, which is not isomorphic to the first.

The question of whether or not Šilov resolutions exist for all contractive Hilbert modules over $R(\Omega^-)$, for Ω an analytic Cauchy domain with more than one hole, has remained elusive, and was first raised in [101].

<u>Problem 3.21</u>. Does every contractive Hilbert module for $R(\Omega^-)$ possess a Šilov resolution?

So far, we have only discussed resolutions by Šilov modules for contractive Hilbert modules. We now focus on the case of Hilbert modules. If $\mathcal{M}$ is a Hilbert module over A, $\mathcal{S}_0, \mathcal{S}_1, \ldots, \mathcal{S}_n$ are Šilov and $\Phi_0, \ldots, \Phi_n$ are bounded module maps such that

$$0 \leftarrow \mathcal{M} \overset{\Phi_0}{\leftarrow} \mathcal{S}_0 \leftarrow \ldots \overset{\Phi_n}{\leftarrow} \mathcal{S}_n \leftarrow 0$$

is exact, then again we always obtain an exact sequence,

$$0 \leftarrow \mathcal{M} \overset{\Phi_0}{\leftarrow} \mathcal{S}_0 \overset{i}{\leftarrow} \tilde{\mathcal{S}}_1 \leftarrow 0$$

where i denotes inclusion, by replacing $\mathcal{S}_1$ by $\tilde{\mathcal{S}}_1 = \ker \Phi_0$ with $\tilde{\mathcal{S}}_1$ Šilov.

Definition 3.22. A Hilbert module $\mathcal{M}$ over A is said to have a quasi-Šilov resolution if there exist Šilov modules $\mathcal{S}_0$ and $\mathcal{S}_1$ for A and bounded maps Φ_0 and Φ_1 with Φ_1 a partial isometry such that the sequence

$$0 \leftarrow \mathcal{M} \overset{\Phi_0}{\leftarrow} \mathcal{S}_0 \overset{\Phi_1}{\leftarrow} \mathcal{S}_1 \leftarrow 0$$

is exact.

Proposition 3.23. A Hilbert module $\mathcal{M}$ over A has a quasi-Šilov resolution if and only if $\mathcal{M}$ is similar to a contractive Hilbert module with a Šilov resolution.

Proof. If $\tilde{\mathcal{M}}$ is a contractive Hilbert module with Šilov resolution, $0 \leftarrow \tilde{\mathcal{M}} \overset{\tilde{\Phi}_0}{\leftarrow} \tilde{\mathcal{S}}_0 \overset{\tilde{\Phi}_1}{\leftarrow} \tilde{\mathcal{S}}_1 \leftarrow 0$ and $S: \tilde{\mathcal{M}} \to \mathcal{M}$ is a bounded, invertible module map, then setting $\mathcal{S}_0 = \tilde{\mathcal{S}}_0$, $\mathcal{S}_1 = \tilde{\mathcal{S}}_1$, $\Phi_0 = S\tilde{\Phi}_0$, $\Phi_1 = \tilde{\Phi}_1$ defines a quasi-Šilov resolution of $\mathcal{M}$.

Conversely, if $\mathcal{M}$ has a quasi-Šilov resolution $0 \leftarrow \mathcal{M} \overset{\Phi_0}{\leftarrow} \mathcal{S}_0 \overset{\Phi_1}{\leftarrow} \mathcal{S}_1 \leftarrow 0$ where we regard $\mathcal{S}_1 \subseteq \mathcal{S}_0$, then let $\tilde{\mathcal{M}}$ be the quotient module of $\mathcal{S}_1$ by $\mathcal{S}_0$. We may regard $\tilde{\mathcal{M}}$ as $\mathcal{S}_1{}^{\perp}$ and we know that $\Phi_0|_{\tilde{\mathcal{M}}}$ is one-to-one and onto. It remains to check that $\Phi_0|_{\tilde{\mathcal{M}}}$ is a module map with respect to the module action on $\tilde{\mathcal{M}}$. Recall that the module action on $\tilde{\mathcal{M}}$ is the compression of the module action on $\mathcal{S}_0$ to $\tilde{\mathcal{M}} \subseteq \mathcal{S}_0$, given by $(f,\tilde{m}) \to P(f\cdot\tilde{m})$ where $P:\mathcal{S}_0 \to \tilde{\mathcal{M}}$ is the orthogonal projection. Hence, $f\cdot\Phi_0(\tilde{m}) = \Phi_0((f,\tilde{m})) + \Phi_0 h = \Phi_0(P(f\cdot\tilde{m}))$, where h is in $\tilde{\mathcal{M}}^{\perp} = \mathcal{S}_1 = \ker \Phi_0$, and so $\Phi_0|_{\tilde{\mathcal{M}}}$ is a module map. □

There is another characterization of those Hilbert modules which have Šilov or, more generally quasi-Šilov resolutions. These characterizations involve some new concepts.

Let A be a function algebra, let $M_n(A)$ denote the set of $n \times n$ matrices with entries from A, that is, $M_n(A) = A \otimes M_n$. We denote a typical element of $M_n(A)$ by (f_{ij}) where f_{ij} is in A, $i,j = 1, \ldots, n$. The set $M_n(A)$ is an algebra in a natural fashion, with the operations given by,

$$\lambda\cdot(f_{ij}) = (\lambda f_{ij}),$$

$$(f_{ij}) + (g_{ij}) = (f_{ij} + g_{ij}),$$

and

$$(f_{ij}) \cdot (g_{ij}) = \left(\sum^n{}_{k=1} f_{ik} \cdot g_{kj}\right).$$

Recall that the operator norm of a matrix (α_{ij}) in M_n is given by,

$$\|(\alpha_{ij})\| = \sup\Big\{ |\sum^n{}_{i,j=1} \alpha_{ij}\lambda_j\bar{\mu}_i| :$$
$$(\lambda_1,\ldots,\lambda_n),(\mu_1,\ldots,\mu_n) \text{ are units vectors in } \mathbb{C}^n \Big\}.$$

For (f_{ij}) in $M_n(A)$ we analogously define,

$$\|(f_{ij})\| = \sup\Big\{\|\textstyle\sum^n_{i,j=1} f_{ij}\lambda_j\bar{\mu}_i\| :$$

$$(\lambda_1,\dots,\lambda_n),(\mu_1,\dots,\mu_n) \text{ are units vectors in } \mathbb{C}^n\Big\}.$$

If $A \subseteq C(X)$, then it is easily seen that this norm is given by,

$$\|(f_{ij})\| = \sup\ \{\|(f_{ij}(x))\| : x \in X\},$$

and consequently, the right hand side of the above equation is independent of the particular space on which we view A as a function algebra.

Now if $\mathcal{M}$ is a bounded Hilbert module, then $\mathcal{M} \otimes \mathbb{C}^n = \mathcal{M}\oplus\dots\oplus\mathcal{M}$ (n copies) is a bounded Hilbert module over $M_n(A)$ with action

$$(f_{ij}) \cdot (h_1\oplus\dots\oplus h_n) = (\textstyle\sum^n_{j=1} f_{ij} \cdot h_j) \oplus\dots\oplus (\textstyle\sum^n_{j=1} f_{nj}\cdot h_j),$$

which is the appropriate analogue of matrix multiplication.

We let $K^n_{\mathcal{M}}$ denote the module bound of $\mathcal{M} \otimes \mathbb{C}^n$ as an $M_n(A)$-module. That is,

$$K^n_{\mathcal{M}} = \sup\ \{\|(f_{ij}) \cdot h\| : \|(f_{ij})\| \le 1 \text{ and } \|h\| \le 1\ \}.$$

An easy argument shows that, $1 \le K_{\mathcal{M}} = K^1_{\mathcal{M}} \le K^2_{\mathcal{M}} \le\dots$, but surprisingly they need not be equal. See Arveson [12] and [87], for examples.

<u>Definition 3.24</u>. A Hilbert module $\mathcal{M}$ over A is <u>completely bounded</u> if $K^\infty_A(\mathcal{M}) = \lim_{n\to\infty} K^n_A(\mathcal{M})$ is finite and <u>completely contractive</u> if $K^n_A(\mathcal{M}) = 1$ for all n.

The following two results explain the relevance of the above definitions.

Theorem 3.2 (Arveson [12]). A necessary and sufficient condition for a Hilbert module to have a Šilov resolution is that it be completely contractive.

Theorem 3.26 (Paulsen [89]). A necessary and sufficient condition for a Hilbert module to have a quasi-Šilov resolution is that it be completely bounded.

Combining the above results with Proposition 3.22, we see that the sets of completely bounded modules, of modules similar to modules with Šilov resolutions, of modules similar to completely contractive modules, and of modules with quasi-Šilov resolutions, all coincide.

In particular, since every contractive Hilbert module for $A(\mathbb{D})$ has a Šilov resolution, a bounded Hilbert module for $A(\mathbb{D})$ will be similar to a contractive Hilbert module if an only if it is completely bounded. Thus, problem 2.4 is equivalent to the following:

Problem 2.4. Is every bounded Hilbert module for $A(\mathbb{D})$ necessarily a completely bounded module?

The necessity of the above conditions in each theorem is fairly straightforward and is a consequence of the following three propositions.

Proposition 3.27. Every contractive Hilbert module over $C(X)$ is completely contractive.

Proof. By Theorem 1.14, we may assume that our Hilbert module is given by a direct integral. In this case, it is straightforward to check that the module is completely contractive. □

Clearly submodules of completely contractive modules are completely contractive and if $\mathcal{M}$ is completely contractive as an A-module, then it is completely contractive as an A_1-module for any $A_1 \subseteq A$. Combining these observations we have that:

Proposition 3.28. Every Šilov module is completely contractive.

Proposition 3.29. If $\mathcal{M}$, $\mathcal{N}$ are bounded Hilbert modules over A and $\Phi:\mathcal{M} \to \mathcal{N}$ is a bounded module map onto $\mathcal{N}$, then $K_A^\infty(\mathcal{N}) \leq \|\Phi\| \cdot K_A^\infty(\mathcal{M})$.

Suppose we replace Šilov modules by submodules for A of contractive Hilbert modules for $C(M_A)$, where M_A is the space of all homomorphisms of A. We shall call such modules subnormal. If instead of considering resolutions by Šilov modules we had considered resolutions by subnormal modules, it is conceivable that we would have obtained a larger class of modules with subnormal resolutions. However, this is not the case.

By Proposition 3.27 and the discussion following, every subnormal module for A is completely contractive. Hence, if $\mathcal{M}$ had a subnormal (respectively, quasi-subnormal) resolution, then $\mathcal{M}$ would be completely contractive (respectively, completely bounded) and so possess a Šilov (respectively, quasi-Šilov) resolution.

In particular, we see that every subnormal module for A possesses a Šilov resolution, and so is the quotient of a Šilov module. This "sweeping to the boundary" is analogous to the fact that a homomorphism in $M_A \backslash \partial A$ can be represented as integration against a measure supported on ∂A.

As we mentioned earlier, it is still an open problem to determine whether or not every contractive Hilbert module over

R(X) for X a compact subset of **C** has a Šilov resolution, even when X is the closure of a finitely connected open subset of **C**. The following result at least contains some information.

Definition 3.30. A function algebra A will be called hypo-Dirichlet if the closure of the linear span of $A + \bar{A}$ is of finite codimension in $C(\partial A)$. We let N(A) denote this codimension.

If $A = R(\Omega^-)$, where Ω is an open finitely-connected set, then A is hypo-Dirichlet [63].

Theorem 3.31 (Douglas-Paulsen [57]). If $\mathcal{M}$ is a contractive Hilbert module for a hypo-Dirichlet algebra A, then $\mathcal{M}$ is completely bounded and $K_A^{\infty}(\mathcal{M}) \leq 2N(A) + 1$.

Corollary 3.32. If $\mathcal{M}$ is a contractive Hilbert module for a hypo-Dirichlet algebra, then $\mathcal{M}$ possesses a quasi-Šilov resolution.

As mentioned earlier, the module viewpoint has contributed little to the above results. Most have come from the study of dilation theory, and the module viewpoint serves largely as a means of organizing this material. What has been gained, however, is the realization that dilation theory is the attempt to analyze Hilbert modules via their resolutions into the simpler Šilov modules. As can be seen from the above results, dilation theory and the classification of Šilov modules is still far from being fully developed, and it is possible that resolutions by some other class of modules would be more fruitful.

Problem 3.33. Are there other "natural" classes of Hilbert modules which lead to a good theory of resolutions?

Unfortunately, most attempts to define classes of Hilbert modules by modifying concepts from commutative algebra, such as freeness, yield either vacuous classes, classes which are far too restrictive to be of value, or return us to the Šilov modules. One fruitful direction, as we shall see in the next chapter, involves the analogues of projective and injective modules.

4 Hypo-projective modules and lifting theorems

Projective modules form the cornerstone for the study of general modules in homological algebra. In this chapter we study an analogue of this notion for Hilbert modules and show that it is related to the lifting theorems in model theory. In particular, a result of Douglas and Foiaş [54] is presented which provides a homological characterization of the Šilov modules for the disk algebra as well as a new proof of the lifting theorem.

<u>Definition</u> <u>4.1</u>. A Hilbert module $\mathcal{P}$ over A is <u>projective</u> if for every pair of bounded Hilbert modules $\mathcal{M}_0$, $\mathcal{M}_1$ over A and every pair of bounded module maps $\Psi: \mathcal{P} \to \mathcal{M}_1$ and $\Phi: \mathcal{M}_0 \to \mathcal{M}_1$, with Φ onto, there exists a bounded module map $\tilde{\Psi}: \mathcal{P} \to \mathcal{M}_0$ such that $\Phi \circ \tilde{\Psi} = \Psi$. The map $\tilde{\Psi}$ is called a <u>lifting</u> of Ψ.

This property is best summarized in terms of the commuting diagram:

$$\begin{array}{ccc} & & \mathcal{M}_0 \\ & \overset{\tilde{\Psi}}{\nearrow} & \downarrow \; \Phi \\ \mathcal{P} & \xrightarrow{\Psi} & \mathcal{M}_1 \\ & & \downarrow \\ & & 0 \end{array}$$

where, as usual, the solid arrows represent the given maps and the broken arrow represents the map that we are asserting exists.

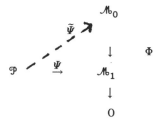

We note that the map $\Phi: \mathcal{M}_0 \to \mathcal{M}_1$ in definition 4.1 can always be replaced by a quotient map without loss of generality, since $\mathcal{M}_1$ is similar to the quotient of $\mathcal{M}_0$ by the kernel of Φ.

Note that the class of projective modules would be vacuous if we attempted to define them strictly within the setting of contractive Hilbert modules, for if we let $\mathcal{M}_0 = \mathcal{M}_1 = \mathcal{P}$, $\Phi(h) = \frac{1}{2}h$ and let Ψ be the identity map, then the only lifting of Ψ would be $\tilde{\Psi}(h) = 2h$, which is not contractive.

In homological algebra, an equally vital role is played by those modules which are universal for the above diagram with the arrows reversed.

<u>Definition</u> <u>4.2</u>. A Hilbert module $\mathcal{I}$ over A is <u>injective</u> if for every pair of Hilbert modules $\mathcal{M}_0$ and $\mathcal{M}_1$ over A and every pair of bounded module maps $\Psi: \mathcal{M}_1 \to \mathcal{I}$, and $\Phi: \mathcal{M}_1 \to \mathcal{M}_0$ with Φ one-to-one and having closed range, there exists a bounded module map $\tilde{\Psi}: \mathcal{M}_0 \to \mathcal{I}$ such that $\Psi = \tilde{\Psi} \circ \Phi$.

In our setting, these two concepts are interchangeable. For a Hilbert module $\mathcal{M}$ over A with $A \subseteq C(X)$, we may also consider $\mathcal{M}$ to be a bounded Hilbert module over $\bar{A}$. If for f in A, $T_f: \mathcal{M} \to \mathcal{M}$ is the bounded operator given by $T_f h = f \cdot h$, then we make $\mathcal{M}$ into a bounded Hilbert $\bar{A}$-module by setting $\bar{f} \cdot h = T_f^* h$. For emphasis, we denote this $\bar{A}$-module by $\mathcal{M}_*$.

<u>Proposition</u> <u>4.3</u>. A Hilbert module $\mathcal{P}$ over A is projective if and only if the Hilbert module $\mathcal{P}_*$ over $\bar{A}$ is injective.

<u>Proof</u>. A simple application of the properties of the adjoint. □

As usual for modules over C(X) little new information is gained, although the following may be of some interest.

Proposition 4.4. Every Hilbert module over C(X) is projective, and hence injective.

Proof. Since every Hilbert module for C(X) is similar to a contractive Hilbert module, it is enough to consider the case where $\mathcal{P}$, $\mathcal{M}_0$ and $\mathcal{M}_1$ are contractive Hilbert modules. But submodules of contractive Hilbert modules for C(X) are reductive, that is, their orthocomplements are also submodules. Thus, any bounded onto module map, $\Phi\colon \mathcal{M}_0 \to \mathcal{M}_1$ will have a splitting, that is, there will exist a bounded module map $\theta\colon \mathcal{M}_1 \to \mathcal{M}_0$ with $\Phi\cdot\theta$ the identity on $\mathcal{M}_1$. Thus, if $\Psi\colon \mathcal{P} \to \mathcal{M}_1$ is any bounded module map, then $\theta\cdot\Psi\colon \mathcal{P} \to \mathcal{M}_0$ will be a lifting of Ψ. Hence, $\mathcal{P}$ is projective, which completes the proof. □

Unfortunately, very little is known about projective Hilbert modules for any other function algebra. In fact, we can not rule out the extreme represented by either of the following at present.

Problem 4.5. Do there exist (non-zero) projective Hilbert modules for every function algebra?

Problem 4.6. Is there any function algebra, other than C(X), with any (non-zero) projective Hilbert modules?

The chief difficulty is that we must consider all of the Hilbert modules, not just the completely bounded Hilbert modules. It is not difficult to show that if we were to consider projective modules in the category of completely bounded Hilbert modules, then such projective modules would be similar to Šilov modules, that is, where all modules in Definition 4.1 are assumed to be Šilov.

Rather than pursuing this, we turn our attention instead to a closely related notion.

Definition 4.7. A contractive Hilbert module $\mathcal{P}$ over A is hypo-projective if for every pair of contractive Hilbert modules $\mathcal{M}_0$ and $\mathcal{M}_1$ over A, such that $\mathcal{M}_1$ is a quotient of $\mathcal{M}_0$ by a partially isometric map Φ, and contractive module map $\Psi: \mathcal{P} \to \mathcal{M}_1$, there exists a contractive module map $\tilde{\Psi}: \mathcal{P} \to \mathcal{M}_0$ with $\Psi = \Phi \circ \tilde{\Psi}$. We shall call such a map $\tilde{\Psi}$ an isometric lifting of Φ.

We begin with a few observations concerning hypo-projective modules. For certain points in the Šilov boundary, we shall see that the module $\mathbb{C}_x$ is hypo-projective.

Definition 4.8. Let A be a function algebra on X. A point x in X will be called a strong separating point for A, provided that the subalgebra generated by the set $\{g \in A \mid g(x) = \|g\|\}$ is dense in A.

It is not hard to see, using the compactness of X, that a strong separating point for A has the property that for every neighborhood V of x there is a function g with $g(x) = \|g\|$ and $|g(y)| < \|g\|$ for y in $X \backslash V$. Thus, every strong separating point is in the Šilov boundary of A.

From this, it follows readily that for the algebras $A(\mathbb{D}^n)$ on $\mathbb{T}^n$, the points in $\mathbb{T}^n$ are the strong separating points. For the algebras $R(\Omega^-)$ on $\partial\Omega$, the points in $\partial\Omega$ are also the strong separating points.

However, it is not the case for a general function algebra A, that every point of ∂A is a strong separating point. For example, let

$X = \{(z,t): z \text{ in } \mathbf{C}, \quad t \text{ in } \mathbb{R}, |z| \leq 1, |t| \leq 1\}$ and let

$A \subseteq C(X)$ consist of those functions f, for which $f(z,0)$ is analytic on $\mathbb{D}\times\{0\}$. Then $\partial A = X$, but no point of the form $(z,0)$, with $|z| < 1$, is a strong separating point for A.

<u>Proposition</u> <u>4.9</u>. If x is a strong separating point for A, then $\mathbb{C}_x$ is a hypo-projective A-module.

<u>Proof</u>. Let $\mathcal{M}_0 \to \mathcal{M}_1 \to 0$ be a quotient and let $\Phi: \mathbb{C}_x \to \mathcal{M}_1$ be a non-zero module map, so that if $h = \Phi(1)$ then $f \cdot h = \Phi(f \cdot 1) = f(x)h$, where $f \cdot h$ is the module action on $\mathcal{M}_1$. Regard $\mathcal{M}_1$ as a subspace of $\mathcal{M}_0$, with the compressed module action, and let $T_g: \mathcal{M}_0 \to \mathcal{M}_0$ be the operator of multiplication by g for g in A, so that

$g(x)h = P_{\mathcal{M}_1}(T_g h)$. If $g(x) = \|g\|$, then the equation $\|g(x)h\| \leq \|P_{\mathcal{M}_1}(T_g h)\| \leq \|T_g h\| \leq \|g\| \, \|h\| \leq |g(x)| \, \|h\|$ implies that $\|P_{\mathcal{M}_1}(T_g h)\| = \|T_g h\|$ and $T_g h = g(x)h$, for all such g. But by hypothesis, these g generate A. Thus, h is also an eigenvector of the module action on $\mathcal{M}_0$.

Hence, defining $\tilde{\Phi}: \mathbb{C}_x \to \mathcal{M}_0$, via $\tilde{\Phi}(\lambda) = \lambda h$ yields the desired isometric lifting of Φ. □

Since finite direct sums of hypo-projective modules are easily seen to be hypo-projective, we see that it is possible to produce many examples of hypo-projective modules.

It is not the case in general, that x in ∂A implies that $\mathbb{C}_x$ is hypo-projective. Let A be the function algebra introduced preceding proposition 4.9 and let $x = (\lambda, 0)$ with $|\lambda| < 1$. If we make $H^2(\mathbb{D})$ a contractive, Hilbert A-module by setting $f \cdot g(z) = f(z,0)g(z)$ for f in A and g in $H^2(\mathbb{D})$, then $\mathbb{C}_x$ is a quotient of $H^2(\mathbb{D})$, but the identity map on $\mathbb{C}_x$ has no lifting to $H^2(\mathbb{D})$.

Proposition 4.10. Every hypo-projective module over A which has a Šilov resolution, is a Šilov module.

Proof. Let $\mathcal{P}$ be a hypo-projective module with a Šilov resolution, so that $\mathcal{P}$ is a quotient of some Šilov module $\mathcal{S}$. If we let $\mathcal{M}_1 = \mathcal{P}$, $\mathcal{M}_0 = \mathcal{S}$ and let Ψ be the identity map in Definition 4.7, then $\tilde{\Psi}: \mathcal{P} \to \mathcal{S}$ must be isometric. Hence $\mathcal{P}$ is a submodule of a Šilov module and so is Šilov. □

By the above result, we see that for the function algebras $A(\mathbb{D})$, $A(\mathbb{D}^2)$, and $R(\mathbb{A})$, all hypo-projective modules are Šilov since in each of these cases all contractive Hilbert modules have Šilov resolutions. For $A(\mathbb{D})$, the converse is also true.

Theorem 4.11 (Douglas-Foias [54]). A contractive Hilbert module for $A(\mathbb{D})$ is hypo-projective if and only if it is Šilov.

Proof. By proposition 4.10, if $\mathcal{S}$ is hypo-projective, then $\mathcal{S}$ is Šilov.

Assume that $\mathcal{S}$ is Šilov, so that multiplication by z on $\mathcal{S}$ is an isometry V which determines $\mathcal{S}$. Let $\mathcal{M} \to \mathcal{M}_1 \to 0$ be a quotient by a partially isometric map, with $\mathcal{M}$ and $\mathcal{M}_1$ contractive Hilbert modules for $A(\mathbb{D})$. These modules are determined by contractions S and T, respectively. Also, let $\Psi: \mathcal{S} \to \mathcal{M}_1$ be a contractive module map. We must prove that there exists a contractive module map $\tilde{\Psi}: \mathcal{S} \to \mathcal{M}$ which lifts Ψ.

If we regard $\mathcal{M} = \mathcal{M}_2 \oplus \mathcal{M}_1$, where $\mathcal{M}_2$ is the kernel of the quotient map, then since $\mathcal{M}_2$ is a submodule we have that, relative to this decomposition of $\mathcal{M}$,

$$S = \begin{pmatrix} R & X \\ 0 & T \end{pmatrix}$$

where R is a contraction operator on $\mathcal{M}_2$, and $X: \mathcal{M}_1 \to \mathcal{M}_2$.

The fact that Ψ is a module map is equivalent to requiring that Ψ intertwines V and T, that is, $\Psi V = T\Psi$. The desired lifting $\tilde{\Psi}: \mathcal{S} \to \mathcal{M}$ must intertwine V and S, $\tilde{\Psi}V = S\tilde{\Psi}$. Since $\tilde{\Psi}$ lifts Ψ, we would have $\tilde{\Psi} = \Psi_2 \oplus \Psi$ where $\Psi_2: \mathcal{S} \to \mathcal{M}_2$.

Performing the operator matrix multiplications, the intertwining condition would be equivalent to

(1) $\Psi_2 V = R\Psi_2 + X\Psi$,

and the condition that $\tilde{\Psi}$ is a contraction to

(2) $\Psi^* \Psi + \Psi^*_2 \Psi_2 \leq 1$.

Thus, we must construct Ψ_2 satisfying these conditions. Assume that $\|S\| < 1$ and $\|T\| < 1$.

We begin with the case that $\dim(\mathcal{M}_2) = 1$ so that $R = \alpha$ for some scalar α with $|\alpha| < 1$.

Equation (1) becomes $\Psi_2(V - \alpha I) = X\Psi$ and since $V - \alpha I$ is one-to-one, we have that necessarily $\Psi_2 = X\Psi(V - \alpha)^{-1}$ on the range of $(V - \alpha)$. If we can show that this expression defines an operator Ψ_2 which is bounded on the range of $(V - \alpha)$, then we can extend to the closure of the range and finally to $\mathcal{S}$ by projecting onto the closure of the range of $(V - \alpha)$.

To accomplish this we show that,

$$(3) \quad 0 \leq \langle (I-\Psi^*\Psi - \Psi^*_2\Psi_2)\, k, k \rangle$$

for $k = (V - \alpha)h$ and an arbitrary h in $\mathcal{G}$. This not only shows that Ψ_2 as defined above is bounded but also that $\Psi_2 \oplus \Psi$ satisfies (2).

Clearly (3) is equivalent to

$$0 \leq \langle ((V-\alpha)^* (V-\alpha) - (V-\alpha)^*\Psi^*\Psi(V-\alpha) - \Psi^*X^*X\,\Psi)h, h \rangle,$$

for all h in $\mathcal{G}$.

The fact that S is a contraction implies that $X = (1 - |\alpha|^2)^{\frac{1}{2}}\, C(1 - T^*T)^{\frac{1}{2}}$ for some contraction C. Combining this fact with the intertwining relations, yields

$$\begin{aligned}(V-\alpha)^*(V-\alpha) - \Psi^*(T-\alpha)^*(T-\alpha)\Psi - (1-|\alpha|^2)\Psi^*(1-T^*T)^{\frac{1}{2}}C^*C\cdot(1-T^*T)^{\frac{1}{2}}\Psi \\ \geq (V-\alpha)^*(V-\alpha) - \Psi^*[(T-\alpha)^*(T-\alpha)-(1-|\alpha|^2)(1-T^*T)]\Psi \\ \geq (V-\alpha)^*(V-\alpha) - \Psi^*(1-\alpha T^*)(1-\bar{\alpha}T)\Psi \\ \geq (V-\alpha)^*(V-\alpha) - \Psi^*(1-\alpha V^*)\Psi^*\Psi(1-\bar{\alpha}V)\Psi \\ \geq (V-\alpha)^*(V-\alpha) - \Psi^*(1-\alpha V^*)(1-\bar{\alpha}V)\Psi \geq 0.\end{aligned}$$

Thus, we have proved that the desired lift $\tilde{\Psi}$ exists when $\dim(\mathcal{M}_2) = 1$.

Now assume that $\dim(\mathcal{M}_2) = n$. Since R is an operator on the finite dimensional space $\mathcal{M}_2$, it will have a chain of invariant subspaces, $0 = \mathcal{M}_2^{(0)} \subseteq \mathcal{M}_2^{(1)} \subseteq \ldots \subseteq \mathcal{M}_2^{(n)} = \mathcal{M}_2$ with $\dim(\mathcal{M}_2^{(i+1)}/\mathcal{M}_2^{(i)}) = 1$. Let $\mathcal{R}_i$ be the quotient of $\mathcal{M}_2 \oplus \mathcal{M}_1$ by the submodule $\mathcal{M}_2^{(i)}$ so that $\mathcal{R}_i = [\mathcal{M}_2 \ominus \mathcal{M}_2^{(i)}] \oplus \mathcal{M}_1$ as a subspace with the compressed action. Then $\mathcal{R}_n = \mathcal{M}_1$, $\mathcal{R}_0 = \mathcal{M}_2 \oplus \mathcal{M}_1$ and $\mathcal{R}_{i+1}$ is a quotient of $\mathcal{R}_i$, with $\dim(\mathcal{R}_i) - \dim(\mathcal{R}_{i+1}) = 1$. Thus, by the above result we may inductively lift $\Psi:\mathcal{S} \to \mathcal{M}_2 = \mathcal{R}_n$ to $\tilde{\Psi}:\mathcal{S} \to \mathcal{R}_0$.

Finally, we must consider the case of a general $\mathcal{M}_2$. For each finite dimensional subspace $\mathcal{F}$ contained in $\mathcal{M}_2$ we set

$$S_{\mathcal{F}} = \begin{pmatrix} R_{\mathcal{F}} & X_{\mathcal{F}} \\ 0 & T \end{pmatrix},$$

on $\mathcal{F} \oplus \mathcal{M}_1$ where $R_{\mathcal{F}} = P_{\mathcal{F}}\, R\, P_{\mathcal{F}}: \mathcal{F} \to \mathcal{F}$, $X_{\mathcal{F}} = P_{\mathcal{F}}X: \mathcal{M}_1 \to \mathcal{F}$ and $P_{\mathcal{F}}$ is the orthogonal projection onto $\mathcal{F}$. By the above arguments there exists a contractive lift $\tilde{\Psi}_{\mathcal{F}} : \mathcal{S} \to \mathcal{F} \oplus \mathcal{M}_1 \subseteq \mathcal{M}_2 \oplus \mathcal{M}_1$ of Ψ. If we partially order these spaces by inclusion and let $\tilde{\Psi}$ be any weak operator limit point of this net of maps, then $\tilde{\Psi}$ will be a contraction, and it is easy to see that $\tilde{\Psi}V = S\tilde{\Psi}$. Thus, $\tilde{\Psi}$ is the desired lift of Ψ. Finally, another limiting procedure allows one to reduce the general case to that where $\|S\| < 1$ and $\|T\| < 1$. □

The above theorem implies a lifting theorem of Sz.-Nagy-Foiaş [101], concerning the existence of liftings of intertwining operators. This was first proved for the case in which the modules are $H^2(\mathbb{D})$ by Sarason [98].

<u>Corollary 4.12</u> <u>(Sz.-Nagy-Foias)</u>. Let $\mathcal{M}_0$ and $\mathcal{M}_1$ be contractive Hilbert modules of $A(\mathbb{D})$, with their unique minimal Šilov resolutions,

$$0 \to \mathcal{S}_0' \to \mathcal{S}_0 \to \mathcal{M}_0 \to 0, \quad 0 \to \mathcal{S}'_1 \to \mathcal{S}_1 \to \mathcal{M}_1 \to 0.$$

Then for every bounded module map $\Phi: \mathcal{M}_0 \to \mathcal{M}_1$ there exist bounded module maps $\tilde{\Phi}: \mathcal{S}_0 \to \mathcal{S}_1$, $\tilde{\Phi}': \mathcal{S}'_0 \to \mathcal{S}'_1$ with $\|\tilde{\Phi}\|$, such that the following diagram commutes:

$$\begin{array}{ccccccc} 0 \leftarrow & \mathcal{S}'_0 & \to & \mathcal{S}_0 & \to & \mathcal{M}_0 & \to 0 \\ & \downarrow\tilde{\Phi}' & & \downarrow\tilde{\Phi} & & \downarrow\Phi & \\ 0 \leftarrow & \mathcal{S}'_1 & \to & \mathcal{S}_1 & \to & \mathcal{M}_1 & \to 0 \end{array}$$

<u>Proof</u>. Since $\mathcal{S}_0$ is hypo-projective, the map from $\mathcal{S}_0$ to $\mathcal{M}_1$ obtained by composing the quotient map from $\mathcal{S}_0$ to $\mathcal{M}_0$ with Φ, lifts to a map $\tilde{\Phi}$ from $\mathcal{S}_0$ to $\mathcal{S}_1$. The map $\tilde{\Phi}'$ is obtained by restricting $\tilde{\Phi}$ to $\mathcal{S}'_0$. □

Using Theorem 2.19, the above Šilov modules $\mathcal{S}_i$ have minimal $C(\mathbb{T})$-extensions $\mathcal{H}_i$, $i = 1$, 2, and hence $\tilde{\Phi}$ lifts to a $C(\mathbb{T})$-module map, $\tilde{\tilde{\Phi}}: \mathcal{H}_1 \rightarrow \mathcal{H}_0$ with $\|\tilde{\tilde{\Phi}}\| = \|\Phi\|$. Since the $A(\mathbb{D})$-modules, $\mathcal{M}_0$ and $\mathcal{M}_1$, are determined by a pair of contractions, C_0 and C_1, and the module map Φ is determined by an intertwining operator X (see Chapter 2) we see that this last result says that X can be lifted to an operator Y, $\|X\| = \|Y\|$, where Y intertwines the unitaries, U_0 and U_1, which determine the $C(\mathbb{T})$-modules $\mathcal{H}_0$ and $\mathcal{H}_1$.

It is not difficult to see that Theorem 4.11 can also be derived as a consequence of the Sz.-Nagy-Foiaş lifting theorem. We do this in Theorem 4.16.

<u>Definition</u> <u>4.13</u>. Let A be a function algebra, $\mathcal{M}_0$ a contractive Hilbert module for A and $\mathcal{S}_0$ a Šilov dominant for $\mathcal{M}_0$. If for every contractive Hilbert A-module $\mathcal{M}_1$ with Šilov dominant $\mathcal{S}_1$, and bounded A-module map $\Phi : \mathcal{M}_0 \rightarrow \mathcal{M}_1$, there exists an A-module map $\tilde{\Phi}: \mathcal{S}_0 \rightarrow \mathcal{S}_1$ with $\|\Phi\| = \|\tilde{\Phi}\|$ such that the diagram

$$
\begin{array}{ccc}
\mathcal{S}_0 & \overset{\tilde{\Phi}}{\rightarrow} & \mathcal{S}_1 \\
\downarrow & & \downarrow \\
\mathcal{M}_0 & \overset{\Phi}{\rightarrow} & \mathcal{M}_1
\end{array}
$$

commutes, then we say that <u>lifting</u> <u>holds</u> <u>for</u> $(\mathcal{S}_0, \mathcal{M}_0)$. If lifting holds for every pair $(\mathcal{S}_0, \mathcal{M}_0)$ where $\mathcal{S}_0$ is a strongly minimal Šilov dominant for $\mathcal{M}_0$, then we say that <u>lifting</u> <u>holds</u> <u>for</u> <u>A</u>.

Let $(\mathcal{S}_0, \mathcal{M}_0)$, $(\mathcal{S}_1, \mathcal{M}_1)$ and Φ be as in definition 4.13 and let $\mathcal{S}'_1$ be the Šilov dominant for $\mathcal{M}_1$ obtained by taking the strongly minimal part of the Šilov resolution determined by $\mathcal{S}_1$. By considering the inclusion of $\mathcal{S}'_1$ into $\mathcal{S}_1$, we see that to determine if lifting holds for $(\mathcal{S}_0, \mathcal{M}_0)$ it is sufficient to consider pairs $(\mathcal{S}'_1, \mathcal{M}_1)$ where $\mathcal{S}'_1$ is a strongly minimal Šilov dominant.

If we required that lifting holds for every pair $(\mathcal{S}_0, \mathcal{M}_0)$ over A, with $\mathcal{S}_0$ either a minimal or weakly minimal Šilov dominant for $\mathcal{M}_0$, then we are putting two apparently stronger conditions on A than given in the above definition. However, it seems likely that these three conditions on A are equivalent. In any case, only the condition given in the above definition, which is the weakest, will play a role in what follows.

<u>Theorem</u> <u>4.14</u>. Let A be a function algebra for which lifting holds. Then every contractive Hilbert A-module which has a Šilov dominant, has a unique, up to unitary equivalence, strongly minimal Šilov dominant.

<u>Proof</u>. Let $\mathcal{M}$ be a contractive Hilbert A-module and let $\mathcal{S}_0$ and $\mathcal{S}_1$ be two strongly minimal Šilov dominants for $\mathcal{M}$. By lifting the identity map on $\mathcal{M}$ we obtain contractive module maps $\Phi_0: \mathcal{S}_0 \to \mathcal{S}_1$ and $\Phi_1: \mathcal{S}_1 \to \mathcal{S}_0$. Let $\Psi = \Phi_1 \circ \Phi_0$ and regard $\mathcal{M}$ as a subspace of $\mathcal{S}_0$ with the compressed action, so that the partial isometry P: $\mathcal{S}_0 \to \mathcal{M}$ simply becomes a projection. Since Ψ is a contraction and $P\Psi = P$, we have that $\Psi(h) = h$ for h in $\mathcal{M}$. But since $A \cdot \mathcal{M}$ is dense in $\mathcal{S}_0$ and Ψ is an A-module map, we have that Ψ is the identity on $\mathcal{S}_0$. Similarly, $\Phi_0 \circ \Phi_1$ is the identity on $\mathcal{S}_1$. □

We have seen earlier that for $A(\mathbb{D})$, $A(\mathbb{D}^2)$, and $R(\mathbb{A})$, every hypo-projective module is necessarily Šilov, and that the converse is true for $A(\mathbb{D})$.

<u>Corollary</u> <u>4.15</u>. For the function algebras $A(\mathbb{D}^2)$ and $R(\mathbb{A})$, lifting does not hold and not every Šilov module is hypo-projective.

<u>Proof</u>. We have given examples (3.9 and 3.20) of strongly minimal Šilov dominants for the point evaluation modules, which are not unitarily equivalent. Thus, lifting does not hold for $A(\mathbb{D}^2)$ or $R(\mathbb{A})$, which is a weaker condition than requiring that the Šilov modules are hypo-projective. □

In fact, from the proof of Theorem 4.14 we see that if $\mathcal{M}$ has two non-unitarily equivalent strongly minimal Šilov dominants, then neither of those Šilov modules can be hypo-projective.

<u>Theorem</u> <u>4.16</u>. Let A be a function algebra for which every contractive Hilbert module has a Šilov resolution and for which lifting holds. Then a contractive Hilbert module for A is Šilov if and only if it is hypo-projective.

<u>Proof</u>. By proposition 4.10, if $\mathcal{S}$ is hypo-projective, then it is Šilov. Assume that $\mathcal{S}$ is Šilov, let $\Phi\colon \mathcal{S} \to \mathcal{M}_0$ be contractive and let $\mathcal{M}_1 \to \mathcal{M}_0 \to 0$ be a quotient. Let $\mathcal{S}_0$ and $\mathcal{S}_1$ be strongly minimal Šilov dominants for $\mathcal{M}_0$ and $\mathcal{M}_1$, respectively. We may regard $\mathcal{M}_0 \subseteq \mathcal{M}_1 \subseteq \mathcal{S}_1$, with the compressed module action. If we let $\mathcal{S}'_0 = A\cdot\mathcal{M}_0 \subseteq \mathcal{S}_1$, then by theorem 4.14, $\mathcal{S}'_0$ and $\mathcal{S}_0$ are unitarily equivalent.

Since lifting holds for A, Φ lifts to $\Phi_0\colon \mathcal{S} \to \mathcal{S}_0$, which when composed with the isomorphism of $\mathcal{S}_0$ and $\mathcal{S}'_0$ yields a module map $\Phi'_0\colon \mathcal{S} \to \mathcal{S}'_0 \subseteq \mathcal{S}_1$. Composing Φ'_0 with the quotient map from $\mathcal{S}_1$ to $\mathcal{M}_1$ yields the desired lifting of Φ. □

Questions about module maps between modules $\Phi\colon \mathcal{M}_0 \to \mathcal{M}_1$ can be reduced to questions about module maps from a module back to itself. Let $\mathcal{M} = \mathcal{M}_0 \oplus \mathcal{M}_1$ with the direct sum action $f\cdot(h_0,h_1) = (f\cdot h_0, f\cdot h_1)$, then the map $\tilde{\Phi}\colon \mathcal{M} \to \mathcal{M}$ given by $\tilde{\Phi}(h_0,h_1) = (0,\Phi h_0)$ is a module map and in fact, $\|\Phi\| = \|\tilde{\Phi}\|$.

Thus, lifting diagrams of the form

$$\begin{array}{ccccc} \mathcal{S}_1 & \to & \mathcal{M}_1 & \to & 0 \\ & & \Phi\downarrow & & \\ \mathcal{S}_0 & \to & \mathcal{M}_0 & \to & 0 \end{array}$$

is equivalent to finding liftings with $\mathcal{S}_1 = \mathcal{S}_0$ and $\mathcal{M}_1 = \mathcal{M}_0$.

Since lifting does not hold for $A(\mathbb{D}^2)$, there exists a contractive Hilbert $A(\mathbb{D}^2)$-module $\mathcal{M}$ with a strongly minimal Šilov dominant $\mathcal{S}$ and a contractive module map $\Phi\colon \mathcal{M} \to \mathcal{M}$ with no contractive lifting to $\mathcal{S}$. The module $\mathcal{M}$ is determined by a pair of commuting contractions T_1 and T_2, and $\mathcal{S}$ is determined by two commuting isometries S_1 and S_2. The map Φ corresponds to a third contraction T_3 which commutes with T_1 and T_2 but has no lift to a contraction S_3 which commutes with S_1 and S_2.

This should be compared with Parrott's example [88] of three commuting contractions with no commuting unitary dilation. Parrott's example yields an $A(\mathbb{D}^2)$-module (given by the first two contractions) such that for <u>any</u> strongly minimal Šilov dominant $\mathcal{S}$ (given by S_1 and S_2) there is no contraction S_3 such that $S_3{}^n$ lifts $T_3{}^n$ for all integers n.

Thus, we see that there is a subtle difference between lifting and dilation. Generally, lifting fixes a Šilov dominant $\mathcal{S}$ and module map $\Phi: \mathcal{M} \to \mathcal{M}$ and seeks a map $\tilde{\Phi}: \mathcal{S} \to \mathcal{S}$ such that $\tilde{\Phi}^n$ lifts Φ^n for all integers n.

The fact that lifting is often enough to imply dilation is somewhat surprising. See [88] for a simple direct argument that the Sz.-Nagy-Foiaş lifting theorem implies Ando's dilation theorem.

J. A. Ball [19] has given an example of a contractive Hilbert $R(\Omega^-)$-module $\mathcal{M}$ with a Šilov dominant $\mathcal{S} \to \mathcal{M} \to 0$ and a bounded, module map $\Phi: \mathcal{M} \to \mathcal{M}$ with no bounded lifting to $\mathcal{S}$. He also proves that Φ will possess a bounded lifting $\tilde{\Phi}$ on $\mathcal{S}$ if the minimal $C(\partial\Omega)$-extension of $\mathcal{S}$, $\mathcal{H} = \int^{\oplus}_{x \in \partial\Omega} \mathcal{H}_x$, satisfies dim $(\mathcal{H}_x) \leq m$ for all x, and some m, in general, $\|\tilde{\Phi}\| \neq \|\Phi\|$.

For similar examples in the $A(\mathbb{D}^2)$ case, one should see D.N. Clark [34], [35].

5 Tensor products of Hilbert modules

In this chapter we introduce the concept of the module tensor product of Hilbert modules. This concept is a useful tool for analyzing Šilov resolutions and for constructing new Hilbert modules. We shall see that exact sequences of Hilbert modules do not necessarily remain exact when each term in the sequence is tensored, as modules, with a fixed Hilbert module. However, with some hypotheses certain portions of an exact sequence are preserved by this operation. The amount that exactness fails can be used to obtain invariants from Šilov resolutions. We begin by recalling tensor product of Hilbert spaces. This module tensor product has been studied earlier in a Banach space setting by J. Taylor [103].

Let $\mathcal{H}$ and $\mathcal{K}$ be Hilbert spaces. We endow their algebraic tensor product with an inner product by setting,

$$\langle h\otimes k,\ h'\otimes k'\rangle = \langle h,\ h'\rangle\ \langle k,\ k'\rangle,$$

and extending linearly. The completion of this inner product space is a Hilbert space which is denoted by $\mathcal{H}\otimes\mathcal{K}$. If $\{e_\alpha\}_{\alpha\epsilon A}$ and $\{f_\beta\}_{\beta\epsilon B}$ are orthonormal bases for $\mathcal{H}$ and $\mathcal{K}$, respectively, then $\{e_\alpha\otimes f_\beta\}_{(\alpha,\beta)\epsilon A\times B}$ is an orthonormal basis for $\mathcal{H}\otimes\mathcal{K}$.

When $L^2(X,\mu)$, $L^2(Y,\nu)$, and $\mathcal{H}$ are separable, then $L^2(X,\mu)\otimes L^2(Y,\nu)$ is isomorphic to $L^2(X\times Y,\mu\times\nu)$ via the map which sends $f(x)\otimes g(y) \to f(x)g(y)$, and $L^2(X,\mu)\otimes\mathcal{H}$ is isomorphic to the space of measurable square-integrable $\mathcal{H}$-valued functions $L^2_{\mathcal{H}}(X,\mu)$ via the map which sends $f(x)\otimes h \to f(x)h$.

If $\mathcal{H}$ and $\mathcal{K}$ are each Hilbert modules for A, then there are two natural ways to make $\mathcal{H}\otimes\mathcal{K}$ a Hilbert module for A. We can define module actions via $f\cdot(h\otimes k) \to (f\cdot h)\otimes k$ or $f\cdot(h\otimes k) \to h\otimes(f\cdot k)$. We call these, respectively, the <u>left</u> and right module tensor products of $\mathcal{H}$ and $\mathcal{K}$ and denote them by ${}_A(\mathcal{H}\otimes\mathcal{K})$ and $(\mathcal{H}\otimes\mathcal{K})_A$, respectively. For x in $\mathcal{H}\otimes\mathcal{K}$ we write $f\cdot x$ and $x\cdot f$ to denote these two actions.

Of course, we are interested in a unique way to define the module action. Note that the closed subspace $\mathcal{N}$ of $\mathcal{H}\otimes\mathcal{K}$ generated by vectors of the form $(f\cdot h)\otimes k - h\otimes(f\cdot k)$ for h in $\mathcal{H}$, k in $\mathcal{K}$, and f in A, is a submodule of both the left and right module tensor products. Thus, we obtain Hilbert modules over A, by forming the quotients ${}_A(\mathcal{H}\otimes\mathcal{K})/\mathcal{N}$ and $(\mathcal{H}\otimes\mathcal{K})_A/\mathcal{N}$. As Hilbert spaces both these modules can be identified as $\mathcal{N}^\perp$ with the module actions given by the compresspions of the, respective, left and right module actions to $\mathcal{N}^\perp$. However, since clearly $P_{\mathcal{N}^\perp}((f\cdot h)\otimes k) = P_{\mathcal{N}^\perp}(h\otimes(f\cdot k))$ we have that both these compressed actions agree. Thus, these quotient modules are isomorphic Hilbert A-modules.

<u>Definition 5.1</u>. Let $\mathcal{H}$ and $\mathcal{K}$ be Hilbert modules for A. Then we denote the quotient module ${}_A(\mathcal{H}\otimes\mathcal{K})/\mathcal{N}$ by $\mathcal{H}\otimes_A\mathcal{K}$ and we call it the <u>tensor product of $\mathcal{H}$ and $\mathcal{K}$ over A</u>. For h in $\mathcal{H}$ and k in $\mathcal{K}$, we let $h\otimes_A k$ denote the image of $h\otimes k$ in $\mathcal{H}\otimes_A\mathcal{K}$.

We begin with some elementary observations.

<u>Proposition 5.2.</u> Let $\mathcal{H}$ and $\mathcal{K}$ be Hilbert modules for A, with $K_A(\mathcal{H})$ and $K_A(\mathcal{K})$ their respective module bounds, then $K_A(\mathcal{H}\otimes_A\mathcal{K}) \le \min\{K_A(\mathcal{H}), K_A(\mathcal{K})\}$. Moreover, if either $\mathcal{H}$ or $\mathcal{K}$ is completely bounded, then $\mathcal{H}\otimes_A\mathcal{K}$ is completely bounded and $K_A^\infty(\mathcal{H}\otimes_A\mathcal{K}) \le \min\{K_A^\infty(\mathcal{H}), K_A^\infty(\mathcal{K})\}$.

<u>Proof</u>. Since $\mathcal{H}\otimes_A\mathcal{K}$ is a quotient of ${}_A\mathcal{H}\otimes\mathcal{K}$, $K_A(\mathcal{H}\otimes\mathcal{K}) \le K_A(\mathcal{H})$. Similarly, $K_A(\mathcal{H}\otimes_A\mathcal{K}) \le K_A(\mathcal{K})$.

The completely bounded case follows in a similar fashion. □

__Proposition 5.3__ Let $\mathcal{H}$ and $\mathcal{K}$ be Hilbert modules for A. If either $\mathcal{H}$ or $\mathcal{K}$ has a Šilov (respectively, quasi-Šilov) resolution, then $\mathcal{H}\otimes_A\mathcal{K}$ has a Šilov (respectively, quasi-Šilov) resolution.

__Proof__. These two statements follow from Theorems 3.25 and 3.26 combined with Proposition 5.2. □

The above result can also be seen directly. If $\mathcal{H}$ has a Šilov resolution, let $\mathcal{S}$ be a Šilov dominant for $\mathcal{H}$ and note that $A(\mathcal{S}\otimes\mathcal{K})$ is a Šilov module. But $_A(\mathcal{H}\otimes\mathcal{K})$ is a quotient of $_A(\mathcal{S}\otimes\mathcal{K})$ and hence $\mathcal{H}\otimes_A\mathcal{K}$ is a quotient of $_A(\mathcal{S}\otimes\mathcal{K})$. Thus, $\mathcal{H}\otimes_A\mathcal{K}$ has a Šilov dominant, and hence a resolution. The remaining cases follow similarly.

We now turn our attention to constructing some examples.

__Example 5.4.__ a) $L^2(\mathbb{T})\otimes_{C(\mathbb{T})}L^2(\mathbb{T}) = (0)$. Let $\{e_n\}_{n\in\mathbb{Z}}$ be the standard orthornormal basis for $L^2(\mathbb{T})$, so that $\{e_n\otimes e_m\}_{(m,n)\in\mathbb{Z}\times\mathbb{Z}}$ is a basis for $L^2(\mathbb{T})\otimes L^2(\mathbb{T})$. Note that $(z^k\cdot e_n)\otimes e_m - e_n\otimes(z^k\cdot e_m) = e_{n+k}\otimes e_m - e_n\otimes e_{m+k}$ belongs to $\mathcal{N}$ for all n,m, and k. Thus, if $f = \sum\alpha_{i,j}\, e_i\otimes e_j$ is orthogonal to $\mathcal{N}$, then

$$0 = \langle f, e_{n+k}\otimes e_m - e_n \otimes e_{m+k}\rangle = \alpha_{n+k,m} - \alpha_{n,m+k}.$$

Hence, $\alpha_{n,m} = \alpha_{n-k,m+k}$ for all k, but since $\|f\|^2 \geq \sum_k|\alpha_{n-k,m+k}|^2$ we must have $\alpha_{n,m} = 0$. Thus, f = 0 and $\mathcal{N}^\perp = (0)$.

b) $L^2(\mathbb{T})\otimes_{A(\mathbb{D})}L^2(\mathbb{T}) = (0)$. The argument is similar. One is led to deduce that $\alpha_{n,m} = \alpha_{n-k,m+k}$ for all $k \geq 0$ which is sufficient to guarantee that
$\alpha_{n,m} = 0$.

c) $L^2(\mathbb{T})\otimes_{A(\mathbb{D})}H^2(\mathbb{D}) = (0)$. Again the argument is similar.

In a moment we shall see that $H^2(\mathbb{D})\otimes_{A(\mathbb{D})}H^2(\mathbb{D}) \neq (0)$. In fact it can be identified with the Bergman module $B^2(\mathbb{D})$, which is the Hilbert space of square-integrable analytic functions on $\mathbb{D}$ with respect to area measure, equipped with the module action given by multiplication. Before we prove this it will be convenient to make some observations about module tensor products over $A(\mathbb{D})$. If $\mathcal{H}$ and $\mathcal{K}$ are Hilbert modules over $R(\Omega^-)$, let T and S denote the operators defined by $Th = z\cdot h$, $Sk = z\cdot k$ on $\mathcal{H}$ and $\mathcal{K}$, respectively.

__Lemma 5.5.__ The subspace $\mathcal{N}$ of $\mathcal{H}\otimes\mathcal{K}$ is the closed span of $\{Th\otimes k - h\otimes Sk: h \in \mathcal{H}, k \in \mathcal{K}\}$.

__Proof__. Note that $T^2h\otimes k-h\otimes S^2k = T(Th)\otimes k-(Th)\otimes Sk+Th\otimes Sk-h\otimes S(Sk)$ which shows that $T^2h\otimes k - h\otimes S^2k$ is the sum of two vectors from the above set. Inductively, one finds that $T^nh\otimes k - h\otimes S^nk$ is the sum of n vectors from the above set. Thus, by linearity, $(p(z)\cdot h)\otimes k - h\otimes(p(z)\cdot k)$ belongs to the span of the above set, for any polynomial p.

Finally, if λ is not in Ω^-, then $(T-\lambda)$ and $(S-\lambda)$ are invertible and

$$(T-\lambda)^{-1}h\otimes k - h\otimes(S-\lambda)^{-1}k = h_1\otimes(S-\lambda)k_1 - (T-\lambda)h_1\otimes k_1 = h_1\otimes Sk_1 - Th_1\otimes k_1,$$

where $h_1 = (T-\lambda)^{-1}h$ and $k_1 = (S-\lambda)^{-1}k$, which is in the above set. Again, inductively $(T-\lambda)^{-n}h\otimes k - h\otimes(S-\lambda)^{-n}k$ can be expressed as a sum of terms in the above set.

Thus, we have that $(f(z)\cdot h)\otimes k - h\otimes(f(z)\cdot k)$ is in the span of the above set for any rational function f. □

Writing $Th\otimes k - h\otimes Sk = (T\otimes I - I\otimes S)(h\otimes\mathcal{K})$ we see that Lemma 5.5 proves that $\mathcal{N}$ is the closure of the range of $(T\otimes I - I\otimes S)$. Alternatively, $\mathcal{N}^\perp = \ker\,(T^*\otimes I - I\otimes S^*)$.

<u>Example 5.6.</u> If $\mathcal{H}$ and $\mathcal{K}$ are Hilbert modules over $A = A(\mathbb{D})$ and $Th = z \cdot h$, $Sk = z \cdot k$ are the operators of multiplication by z on $\mathcal{H}$ and $\mathcal{K}$, respectively, then we let $R = T \otimes_A I = I \otimes_A S$ denote the operator of multiplication by z on $\mathcal{H} \otimes_A \mathcal{K}$.

To illustrate this construction, we shall study the operator R when T and S are contractive weighted shifts.

Let S_α and S_β be weighted shifts on l^2 defined by $S_\alpha e_i = \alpha_{i+1} e_{i+1}$, $S_\beta e_i = \beta_{i+1} e_{i+1}$, with $\alpha_i > 0$ and $\beta_i > 0$ for all i, where $\{e_i\}_{i=0}^{\infty}$ denotes the canonical basis for l^2. If we require $|\alpha_i|$, $|\beta_i| \leq 1$, for all i, then S_α and S_β are contractions and determine two $A(\mathbb{D})$ module structures for l^2. We shall show that R is unitarily equivalent to another weighted shift on l^2.

By Lemma 5.5, $h = \sum \delta_{ij} e_i \otimes e_j$ will belong to $\mathcal{N}^\perp$ if and only if for all k and m,

$$0 = \langle h, S_\alpha e_k \otimes e_m - e_k \otimes S_\beta e_m \rangle = \alpha_{k+1} \delta_{k+1,m} - \beta_{m+1} \delta_{k,m+1}.$$

Note that if we choose $\{\delta_{n,0}\}_{n=0}^{\alpha}$ then the remaining $\delta_{k,m}$'s are determined. Thus, if we set $w_i = \prod_{l=1}^{i} \alpha_l$, $v_i = \prod_{l=1}^{i} \beta_l$, $v_o = w_o = 1$, and let $h_n = \sum_{i=0}^{n} w_{n-i}^{-1} v_i^{-1} e_{n-i} \otimes f_i$, $n = 0,1,2,\ldots$, then the span of the h_n's are dense in $\mathcal{N}^\perp$. Moreover, since $\langle h_n, h_m \rangle = 0$ for $m \neq n$, $\{h_n/\|h_n\|\}_{n=0}^{\infty}$ is an orthonormal basis for $\mathcal{N}^\perp$.

Since $R = P_{\mathcal{N}^\perp}(S_\alpha \otimes I)|_{\mathcal{N}^\perp}$ we have that $\langle Rh_n, h_k \rangle = \langle (S_\alpha \otimes I)h_n, h_k \rangle = 0$ if $k \neq n + 1$, while $\langle (S_\alpha \otimes I)h_n, h_k \rangle = \sum_{i=0}^{n} \alpha_{n+1-i} w_{n-i}^{-1} v_i^{-1} = \|h_n\|^2$.

Thus, R is a weighted shift with weight sequence

$\|h_n\|/\|h_{n+1}\|$.

Note that when $S_\alpha = S_\beta = S$, the standard unilateral shift, that is, $\alpha_i = \beta_i = 1$ for all i, then this weight sequence is $(n/n+1)^{\frac{1}{2}}$, which is the weight sequence for the operator of multiplication by z on the Bergman space, $B^2(\mathbb{D})$. Thus, $H^2(\mathbb{D}) \otimes_A H^2(\mathbb{D})$ is isomorphic to $B^2(\mathbb{D})$ as A-modules.

Both the unilateral shift and the Bergman shift operator are subnormal, which suggests the following:

<u>Problem</u> <u>5.7.</u> If S_α and S_β are subnormal weighted shifts, then is R subnormal?

T. Trent has recently shown that the answer is no.

The corresponding problem with "subnormal" replaced by "hyponormal" has recently been solved, in the affirmative, by Badri and Szeptycki [17]. More general results along the above lines have been obtained by Salinas and Badri [15], [16] and [95]. They consider Hilbert spaces of analytic functions on an analytic Cauchy domain Ω, which have reproducing kernels. Under certain conditions, they prove that the tensor product of two such spaces over $R(\Omega^-)$ is isomorphic to a Hilbert space of analytic functions on Ω whose reproducing kernel is the product of the two original kernels. For example, $H^2(\mathbb{D})$ has reproducing kernel $k(z,w) = (1-\overline{w}z)^{-1}$ and it is known that the square of this function is the reproducing kernel for $B^2(\mathbb{D})$, so that their results yield an alternate proof of the fact that $H^2(\mathbb{D}) \otimes_{A(\mathbb{D})} H^2(\mathbb{D})$ and $B^2(\mathbb{D})$ are isomorphic.

The construction of the module tensor product has appeared in several other places in the recent literature embedded in various proofs. One instance of this is in generalizations of Rota's model theorem ([93], [74], and [105]). A careful reading of Herrero [75,Theorem 3.33] shows that what is actually proved is that if T is an operator on $\mathcal{H}$ whose spectrum contained in an analytic Cauchy domain Ω, then T is similar to $M_z \otimes_{R(\Omega^-)} I$, where M_z is the operator of multiplication by z on $H^2(\partial\Omega, ds)$ and ds represents arc-length measure. We explain this observation in more detail below.

__Example 5.8.__ Let Ω be an analytic Cauchy domain and let T be an operator on a Hilbert space $\mathcal{H}$ whose spectrum is contained in Ω. We regard $\mathcal{H}$ as a Hilbert $R(\Omega^-)$-module with action $f \cdot h = f(T)h$, for f in $R(\Omega^-)$ and h in $\mathcal{H}$. Let $H^2(\partial\Omega, ds)$ be the Hilbert space of boundary values of square-integrable analytic functions on Ω with respect to arc-length measure, ds. This is a contractive, Hilbert $R(\Omega^-)$-module with action given by multiplication, $(f \cdot g)(z) = f(z)g(z)$, for f in $R(\Omega^-)$ and g in $H^2(\partial\Omega, ds)$. Note also that if we let M_z denote the operator of multiplication by z, then $f \cdot g = f(M_z)g$.

The Hilbert space $H^2(\partial\Omega, ds) \otimes \mathcal{H}$ can be identified with the Hilbert space $H^2_{\mathcal{H}}(\partial\Omega, ds)$ of boundary values of square-integrable $\mathcal{H}$-valued analytic functions on Ω with respect to arc-length measure. We identify $g \otimes h$ with the $\mathcal{H}$-valued analytic function $g(z)h$.

With this identification the left and right module actions on $H^2(\partial\Omega, ds) \otimes \mathcal{H}$ become the module actions on $H^2_{\mathcal{H}}(\partial\Omega, ds)$ given, respectively, by $(f \cdot k)(z) = f(z)k(z)$ and $(k \cdot f)(z) = f(T)k(z)$. We see that the subspace $\mathcal{N}$ of $H^2(\partial\Omega, ds) \otimes \mathcal{H}$, whose orthogonal complement yields the module tensor product, is , by Lemma 5.5, identified with the range of the operator of multiplication by $(z-T)$ on $H^2_{\mathcal{H}}(\partial\Omega, ds)$.

We wish to show that the operator T on $\mathcal{H}$ is similar to the operator $B = M_z \otimes_{R(\Omega^-)} I$ on $H^2(\partial\Omega, ds) \otimes_{R(\Omega^-)} \mathcal{H}$. Recall that B is equal to both the left and right module action of multiplication by z compressed to $\mathcal{N}^\perp$. That is,
$Bk = P(Tk) = P(\mathcal{M}_z k)$, where k is in $\mathcal{N}^\perp$ and P denotes the orthogonal projection from $H^2_{\mathcal{H}}(\partial\Omega, ds)$ onto $\mathcal{N}^\perp$.

Identifying $\mathcal{H}$ with the constant functions in $H^2_{\mathcal{H}}(\partial\Omega,ds)$, let R: $\mathcal{H}\to\mathcal{N}^{\perp}$ be the restriction of P to $\mathcal{H}$. Then for h in $\mathcal{H}$, RTh = PTh = PTPh, since $\mathcal{H}$ is invariant under multiplication by T. But PTh = BPh = BRh, so we have that RT = BR. Thus, to show that T and B are similar it will be enough to show that R is invertible.

To this end consider the bounded operator L: $H^2_{\mathcal{H}}(\partial\Omega,ds)\to\mathcal{H}$ given by

$$Lk = \frac{1}{2\pi i}\int_{\partial\Omega} (w-T)^{-1}k(w)\ dw.$$

Note that,

$$k(z) - Lk = \frac{1}{2\pi i}\int_{\partial\Omega} [(w-z)^{-1} - (w-T)^{-1}]k(w)\ dw$$

$$= (z-T)\ \frac{1}{2\pi i}\int_{\partial\Omega} [(w-T)^{-1}(w-z)^{-1}]k(w)\ dw = (z-T)\ k_1(z),$$

where $k_1(z)$ is in $H^2_{\mathcal{H}}(\partial\Omega,ds)$. Therefore (k - Lk) lies in $\mathcal{N}$, and Pk = PLk = RLk, for any k in $H^2_{\mathcal{H}}(\partial\Omega,ds)$. Thus, for any k in $\mathcal{N}^{\perp}$, k = Pk = R(Lk) and we see that R is onto.

To see that R is invertible it remains to show that R is one-to-one. To this end note that for h in $\mathcal{H}$, Lh = h, while if $k(z) = (z-T)\ k_1(z)$ for $k_1(z)$ in $H^2_{\mathcal{H}}(\partial\Omega,ds)$, then

$$Lk = \frac{1}{2\pi i}\int_{\partial\Omega} k_1(w)\ dw = 0.$$

Thus, $\mathcal{N}$ is contained in the kernel of L. Since the kernel of R is $\mathcal{H}\cap\mathcal{N}$ we have that any vector h in the kernel of R satisfies both Lh = 0 and Lh = h. Thus, R is one-to-one.

Since we have that T and B are similar operators, this implies that $\mathcal{H}$ and $H^2(\partial\Omega,ds)\otimes_{R(\Omega^-)}\mathcal{H}$ are similar as Hilbert $R(\Omega^-)$-modules. Also, we have that $H^2(\partial\Omega,ds)\otimes_{R(\Omega^-)}\mathcal{H}$ is a quotient of the module $H^2(\partial\Omega,ds)\otimes\mathcal{H}$ with its left module action, and this latter module is clearly Šilov. Thus, the Hilbert $R(\Omega^-)$-module $\mathcal{H}$, has a quasi-Šilov resolution.

From the above arguments we see that anytime a Hilbert $R(\Omega^-)$-module $\mathcal{H}$ has the property that the operator T of multiplication by z on $\mathcal{H}$ has spectrum inside the open set Ω, then $\mathcal{H}$ has a quasi-Šilov resolution.

In the remainder of this chapter we focus on how module maps and exact sequences behave with respect to the module tensor product. In particular, we are interested in the information that can be obtained by tensoring Šilov resolutions with some elementary modules, such as $\mathbb{C}_x$, for x in the maximal ideal space of A.

We begin by constructing tensor products of module maps.

<u>Proposition</u> <u>5.9.</u> If $\mathcal{L}_1$ and $\mathcal{M}_1$ are bounded Hilbert modules for A, and $\mathcal{L}_0$ and $\mathcal{M}_0$ are submodules of $\mathcal{L}_1$ and $\mathcal{M}_1$, respectively, then any bounded module map Φ: $\mathcal{L}_1 \to \mathcal{M}_1$ with $\Phi(\mathcal{L}_0)$ contained in $\mathcal{M}_0$, induces a unique bounded module map Φ: $\mathcal{L}_1/\mathcal{L}_0 \to \mathcal{M}_1/\mathcal{M}_0$ such that the following diagram commutes:

$$\begin{array}{ccccccccc} 0 & \to & \mathcal{L}_0 & \to & \mathcal{L}_1 & \to & \mathcal{L}_1/\mathcal{L}_0 & \to & 0 \\ & & \Phi\downarrow & & \Phi\downarrow & & \overline{\Phi}\downarrow & & \\ 0 & \to & \mathcal{M}_0 & \to & \mathcal{M}_1 & \to & \mathcal{M}_1/\mathcal{M}_0 & \to & 0. \end{array}$$

Moreover, we have that $\|\overline{\Phi}\| \leq \|\Phi\|$.

<u>Proof</u>. A simple diagram chase. □

<u>Proposition</u> <u>5.10.</u> Let $\mathcal{L}_i$ and $\mathcal{M}_i$ be bounded Hilbert modules for A, and let Φ_i: $\mathcal{L}_i \to \mathcal{M}_i$ be bounded module maps, i = 1, 2. Then there is a bounded module map,

$$\Phi_1 \otimes_A \Phi_2\colon \ \mathcal{L}_1 \otimes_A \mathcal{L}_2 \to \mathcal{M}_1 \otimes_A \mathcal{M}_2 \text{ with}$$

$$\Phi_1 \otimes_A \Phi_2 (h_1 \otimes_A h_2) = \Phi_1(h_1) \otimes_A \Phi_2(h_2) \text{ and}$$

$$\|\Phi_1 \otimes_A \Phi_2\| \leq \|\Phi_1\| \cdot \|\Phi_2\|.$$

Moreover, if Φ_1 and Φ_2 are onto, then $\Phi_1 \otimes_A \Phi_2$ is onto.

Proof. Let $\mathcal{N}_0$ and $\mathcal{N}_1$ denote the respective submodules of ${}_A(\mathcal{L}_0 \otimes \mathcal{L}_1)$ and ${}_A(\mathcal{M}_0 \otimes \mathcal{M}_1)$, that are the kernels of the quotient maps onto $\mathcal{L}_0 \otimes_A \mathcal{L}_1$ and $\mathcal{M}_0 \otimes_A \mathcal{M}_1$. By the above proposition it is enough to note that the bounded operator $\Phi_1 \otimes \Phi_2 : \mathcal{L}_1 \otimes \mathcal{L}_2 \to \mathcal{M}_1 \otimes \mathcal{M}_2$ is an A-module map for this left action and maps $\mathcal{N}_0$ into $\mathcal{N}_1$. □

It should be noted that the notation introduced in Propositon 5.10 is consistent with the notation introduced in Example 5.6. Setting $\mathcal{L}_1 = \mathcal{L}_2 = \mathcal{H}$, $\mathcal{M}_1 = \mathcal{M}_2 = \mathcal{K}$, $\Phi_1 = T$, and $\Phi_2 = I$, we see that the operator $T \otimes_A I$ of Example 5.6 coincides with the operator $\Phi_1 \otimes_A \Phi_2$ of Proposition 5.10. The construction in Proposition 5.10 generalizes the construction of the operator $T \otimes_A I$.

We now study the behavior of exact sequences when we tensor each term of the sequence with a fixed bounded Hilbert module and tensor each map in the sequence with the identity map. In general, the new sequence of maps will no longer be exact, but this lack of exactness yields new information about the sequence.

We shall need the following lemma.

Lemma 5.11. Let $\mathcal{M}_1$ and $\mathcal{M}_2$ be Hilbert modules for A with $\text{rank}_A(\mathcal{M}_1)$ finite and the dimension of $\mathcal{M}_2$ finite. Then the dimension of $\mathcal{M}_1 \otimes_A \mathcal{M}_2$ is finite and satisfies
$\dim(\mathcal{M}_1 \otimes_A \mathcal{M}_2) \leq \text{rank}_A(\mathcal{M}_1) \cdot \dim(\mathcal{M}_2)$.

Proof. Let $\text{rank}_A(\mathcal{M}_1) = n$ and let $h_1, \ldots, h_n$ be a generating set for $\mathcal{M}_1$ over A. Also, let $k_1, \ldots, k_m$ be a basis for $\mathcal{M}_2$. We claim that $\{h_i \otimes_A k_j\}_{i=1\,j=1}^{n,m}$ is a spanning set for $\mathcal{M}_1 \otimes_A \mathcal{M}_2$. To see this, note that $\{h_i \otimes k_j\}$ is clearly a generating set in ${}_A(\mathcal{M}_1 \otimes \mathcal{M}_2)$. But for each φ in A there are scalars α_{ij} such that $\varphi \cdot k_j = \sum_{i=1}^{m} \alpha_{ij} k_i$. Therefore, $(\varphi \cdot h_p) \otimes_A k_j = h_p \otimes_A \varphi \cdot k_j = \sum_{i=1}^{m} \alpha_{ij}\, h_p \otimes_A k_i$ and so we see that the span of $\{h_i \otimes_A k_j\}$ is dense in $\mathcal{M}_1 \otimes_A \mathcal{M}_2$ and hence spans.□

Recall that a sequence of maps $\mathcal{M}_2 \xrightarrow{\Phi_1} \mathcal{M}_1 \xrightarrow{\Phi_0} \mathcal{M}_0$ is <u>exact</u> at $\mathcal{M}_1$ if the range of Φ_1 is the kernel of Φ_0.

<u>Theorem 5.12.</u> Let $\mathcal{M}_1$ be a Hilbert module over A with $\text{rank}_A(\mathcal{M}_1)$ finite and let $\mathcal{M}_2 \xrightarrow{\Phi_1} \mathcal{M}_1 \xrightarrow{\Phi_0} \mathcal{M}_0 \to 0$ be an exact sequence of Hilbert modules for A. If $\mathcal{L}$ is a finite dimensional Hilbert module for A, then the sequence of Hilbert modules,

$$\mathcal{M}_2 \otimes_A \mathcal{L} \xrightarrow{\Phi_1 \otimes_A 1} \mathcal{M}_1 \otimes_A \mathcal{L} \xrightarrow{\Phi_0 \otimes_A 1} \mathcal{M}_0 \otimes_A \mathcal{L} \to 0$$

is exact.

<u>Proof</u>. Consider the following commuting diagram, whose columns are exact and whose middle row is exact,

$$\begin{array}{ccccccc}
 & & 0 & & 0 & & 0 & & \\
 & & \downarrow & & \downarrow & & \downarrow & & \\
 & & \mathcal{N}_2 & \longrightarrow & \mathcal{N}_1 & \longrightarrow & \mathcal{N}_0 & & \\
 & & \downarrow & & \downarrow & & \downarrow & & \\
0 & \longrightarrow & \mathcal{M}_2 \otimes \mathcal{L} & \xrightarrow{\Phi_1 \otimes 1} & \mathcal{M}_1 \otimes \mathcal{L} & \xrightarrow{\Phi_0 \otimes 1} & \mathcal{M}_0 \otimes \mathcal{L} & \longrightarrow & 0 \\
 & & \downarrow & & \downarrow & & \downarrow & & \\
 & & \mathcal{M}_2 \otimes_A \mathcal{L} & \xrightarrow{\Phi_1 \otimes_A 1} & \mathcal{M}_1 \otimes_A \mathcal{L} & \xrightarrow{\Phi_0 \otimes_A 1} & \mathcal{M}_0 \otimes_A \mathcal{L} & & \\
 & & \downarrow & & \downarrow & & \downarrow & & \\
 & & 0 & & 0 & & 0. & &
\end{array}$$

We see that $\Phi_0 \otimes_A 1$ has dense range and so by Lemma 5.11 is onto. To check exactness at $\mathcal{M}_1 \otimes_A \mathcal{L}$, let u be in $\mathcal{M}_1 \otimes_A \mathcal{L}$ with

$\Phi_0 \otimes_A 1(u) = 0$. If we lift u to v in $\mathcal{M}_1 \otimes \mathcal{L}$, then $\Phi_0 \otimes 1(v)$ must be in $\mathcal{N}_0$. Since $\mathcal{N}_0$ is spanned by vectors of the form $\sum[(f_i \cdot h_i) \otimes k_i - h_i \otimes (f_i \cdot k_i)]$, if we choose $\tilde{h}_i$ in $\mathcal{M}_1$ with $\Phi_0(\tilde{h}_i) - h_i$ small, then $\sum[(f_i \cdot \tilde{h}_i) \otimes k_i - \tilde{h}_i \otimes (f_i \cdot k_i)]$ is in $\mathcal{N}_1$ and almost maps onto the previous vector. Thus, $\Phi_0 \otimes 1(\mathcal{N}_1)$ is dense in $\mathcal{N}_0$. Hence, we can choose v' in $\mathcal{N}_1$ such that the norm of $\Phi_0 \otimes 1(v+v')$ is arbitrarily small.

However, by the exactness of the middle row, we can choose w in $\mathcal{M}_2 \otimes \mathcal{L}$ such that the norm of $[\Phi_1 \otimes 1(w) - v - v']$ is arbitrarily small. If $\overline{w}$ denotes the image of w in $\mathcal{M}_2 \otimes_A \mathcal{L}$, then the norm of $[\Phi_1 \otimes_A 1(\overline{w}) - u]$ is arbitrarily small.

Thus, u belongs to the closure of the range of $\Phi_1 \otimes_A 1$. But by Lemma 5.11, $\mathcal{M}_1 \otimes_A \mathcal{L}$ is finite dimensional, and so u is actually in the range of $\Phi_1 \otimes_A 1$, from which it follows that we have exactness at $\mathcal{M}_1 \otimes_A \mathcal{L}$.

This completes the proof. □

We show in example 5.15 that even with the hypotheses of the above theorem, injectivity of Φ_1 does not imply injectivity of $\Phi_1 \otimes_A 1$. Thus, the exact sequence of theorem 5.12 can not be extended one term to the left.

A necessary condition for exactness at $\mathcal{M}_1 \otimes_A \mathcal{L}$ in theorem 5.12, is that the range of $\Phi_1 \otimes_A 1$ be closed. The above proof shows that this condition is also sufficient. Following proposition 5.14, we give an example which shows that when $\text{rank}_A(\mathcal{M}_1)$ is not finite then it is possible for the range of $\Phi_1 \otimes_A 1$ to be non-closed, even when $\mathcal{L}$ is finite dimensional. This shows that the hypothesis that $\text{rank}_A(\mathcal{M}_1)$ is finite is essential to theorem 5.12.

<u>Problem</u> <u>5.13.</u> Find hypotheses on an A-module map $\Phi : \mathcal{M}_2 \rightarrow \mathcal{M}_1$ such that $\Phi_1 \otimes_A 1: \mathcal{M}_2 \otimes_A \mathcal{L} \rightarrow \mathcal{M}_1 \otimes_A \mathcal{L}$ has closed range for any Hilbert A-module $\mathcal{L}$. Can this be done under the additional hypotheses that $\mathcal{L}$ is finite dimensional or finite

rank?

Before studying any applications of the above theorem it will be convenient to take a closer look at the result of tensoring with the modules of the form C_X. In fact, we shall consider an arbitrary finite dimensional cyclic module.

__Theorem 5.14__. Let $\mathcal{L}$ be a bounded, finite dimensional Hilbert A-module with a cyclic vector e_0, and let $J = \{f \in A: f \cdot e_0 = 0\}$. If $\mathcal{M}$ is a bounded Hilbert A-module, then $\mathcal{M} \otimes_A \mathcal{L}$ and the quotient module, $\mathcal{M}/[J\cdot\mathcal{M}]^-$ are similar A-modules.

__Proof.__ Define a map ϕ: $\mathcal{M} \otimes_A \mathcal{L} \rightarrow \mathcal{M}/[J\cdot\mathcal{M}]^-$ by setting $\phi(h \otimes f \cdot e_0) = f \cdot h + [J\cdot\mathcal{M}]^-$, and extending linearly. We leave it to the reader to check that this map is well-defined.

To see that ϕ is bounded, choose $f_1, \ldots, f_n$ in A such that $\{f_i \cdot e_0\}_{i=1}^n$ is an orthonormal basis for $\mathcal{L}$. Writing an arbitrary element of $\mathcal{M} \otimes \mathcal{L}$ as $\sum_{i=1}^n h_i \otimes f_i \cdot e_0$, we have that,

$$\|\phi(\textstyle\sum_{i=1}^n h_i \otimes f_i \cdot e_0)\| = \|\textstyle\sum_{i=1}^n f_i \cdot h_i\|$$

$$\leq K_{\mathcal{M}} \cdot \textstyle\sum_{i=1}^n \|f_i\| \cdot \|h_i\| \leq K_{\mathcal{M}} (\textstyle\sum_{i=1}^n \|f_i\|^2)^{\frac{1}{2}} (\textstyle\sum_{i=1}^n \|h_i\|^2)^{\frac{1}{2}}$$

$$\leq K_{\mathcal{M}} (\textstyle\sum_{i=1}^n \|f_i\|^2)^{\frac{1}{2}} \|\textstyle\sum_{i=1}^n h_i \otimes f_i \cdot e_0\|,$$

and so ϕ is bounded.

If we let $\mathcal{N}$ denote the kernel of the quotient map from $\mathcal{M} \otimes \mathcal{L}$ to $\mathcal{M} \otimes_A \mathcal{L}$, then it is clear that $\mathcal{N}$ is contained in the kernel of ϕ. Hence, there is a bounded induced map, $\hat{\phi}: \mathcal{M} \otimes_A \mathcal{L} \rightarrow \mathcal{M}/[J\cdot\mathcal{M}]^-$.

Clearly, ϕ is a left A-module map, and so ϕ is an A-module map. Since $\phi(h \otimes e_0) = h + [J\cdot\mathcal{M}]^-$, ϕ and hence $\hat{\phi}$ is onto. To complete the proof of the theorem it will suffice to show that $\hat{\phi}$ is one-to-one.

To this end define $\psi: \mathcal{M} \to \mathcal{M}\otimes_A \mathcal{L}$ via $\psi(h) = h\otimes_A e_0$. Then ψ is bounded, and $J\cdot\mathcal{M}$ is contained in the kernel of ψ_0 since $\psi(f\cdot h) = (f\cdot h) \otimes_A e_0 = h \otimes_A (f\cdot e_0)$.

Thus, there is an induced map, $\hat{\psi}: \mathcal{M}/[J\cdot\mathcal{M}]^- \to \mathcal{M}\otimes_A \mathcal{L}$. It is easily checked that $\hat{\psi} \circ \hat{\phi}$ is the identity on $\mathcal{M}\otimes_A \mathcal{L}$ and hence $\hat{\phi}$ must be one-to-one, which completes the proof of the theorem. □

Note that the above set J is always a closed ideal in A. We call it the <u>annihilating</u> <u>ideal</u> of $\mathcal{L}$.

Applying the theorem to $\mathbb{C}_{x_0}$, which has cyclic vector 1, and annihilating ideal $I_{x_0} = \{f \in A : f(x_0) = 0\}$, we see that $\mathcal{M} \otimes_A \mathbb{C}_{x_0}$ is similar to $\mathcal{M}/[I_{x_0}\mathcal{M}]^-$.

We remark that if A is singly-generated with generator z, and $z\cdot h = Th$, then $[I_{x_0}\cdot\mathcal{M}]^- = \ker(T^* - \overline{z(x_0)})^{\perp}$ and $\mathcal{M}\otimes_A \mathbb{C}_{x_0}$ is isomorphic to $\ker(T^* - \overline{z(x_0)})$ with the compressed module action.

Also, note that $\mathcal{M}\otimes_A \mathbb{C}_{x_0}$ is always isomorphic to the module direct sum of copies of $\mathbb{C}_{x_0}$, so that the only invariant for these modules is $\dim(\mathcal{M}\otimes_A \mathbb{C}_{x_0})$, which is $\dim(\ker(T^* - \overline{z(x_0)}))$ in the above case.

We now give an example which shows that the hypothesis that $\mathrm{rank}_A(\mathcal{M}_1)$ is finite was necessary in theorem 5.12. In fact, we shall show that without this hypothesis it is possible to have an inclusion of contractive Hilbert A-modules $0 \to \mathcal{M}_2 \xrightarrow{\Phi} \mathcal{M}_1$ and a 1-dimensional Hilbert A-module $\mathcal{L}$ such that $\Phi\otimes_A 1: \mathcal{M}_2\otimes_A \mathcal{L} \to \mathcal{M}_1\otimes_A \mathcal{L}$ does not even have closed range.

To this end, let $A = A(\mathbb{D})$, and let $\mathcal{M}_1 = H^2(\mathbb{D}^2)$ with action $f\cdot g = f(z)g(z,w)$ for f in A and g in $\mathcal{M}_1$. We let $\mathcal{M}_2$ be the (closed) submodule of $\mathcal{M}_1$ consisting of functions satisfying $g(z,2z) = 0$, when $|z| \le \frac{1}{2}$, and let Φ denote the inclusion map.

If $\mathbb{C}_0$ denotes the 1-dimensional Hilbert $A(\mathbb{D})$-module with action $f\cdot\lambda = f(0)\lambda$ for f in $A(\mathbb{D})$ and λ in $\mathbb{C}$, then by the remarks following proposition 5.14, $H^2(\mathbb{D}^2)\otimes_A \mathbb{C}_0$ is isomorphic to the kernel of M_z^*, where M_z is the operator of multiplication by z. This space is easily seen to be the submodule of $H^2(\mathbb{D}^2)$ consisting of functions which are constant in the z-variable, which we may identify with $H^2(\mathbb{D})$ in the w-variable.

With these identifications, the range of $\Phi\otimes_A 1$ is easily seen to be the orthogonal projection of $\mathcal{M}_2$ onto the space of functions which are constant in the z-variable. Clearly, $(2z)^n - w^n$ is the function in $\mathcal{M}_2$ of minimal norm which projects onto w^n. Thus, the range of $\Phi\otimes_A 1$ will be dense in $H^2(\mathbb{D}^2)\otimes_A \mathbb{C}_0$, but will only contain functions $g(w) = \sum_{n=0}^{\infty}\alpha_n w^n$, for which $\sum_{n=0}^{\infty} 4^n|\alpha_n|^2$ is finite.

We now turn our attention to an example where the hypotheses of theorem 5.12 are met.

<u>Example 5.15</u>. Let θ be an inner function on $\mathbb{D}$ and let $\mathcal{M}_\theta$ denote the quotient contractive Hilbert module for $A(\mathbb{D})$, $\mathcal{M}_\theta = H^2(\mathbb{D})/\theta\cdot H^2(\mathbb{D})$.

By the above remarks it is not too difficult to see that $\mathcal{M}_\theta\otimes_{A(\mathbb{D})}\mathbb{C}_z$ is isomorphic to $\mathbb{C}_z$ when $\theta(z) = 0$ and $|z| < 1$ and is 0 otherwise. We wish to point out how this can be seen by theorem 5.12.

If we consider the exact sequence,

$$0\rightarrow H^2(\mathbb{D}) \xrightarrow{\theta} H^2(\mathbb{D}) \rightarrow \mathcal{M}_\theta \rightarrow 0,$$

then by Theorem 5.12, we obtain an exact sequence,

$$H^2(\mathbb{D})\otimes_A \mathbb{C}_z \xrightarrow{\theta\otimes_A 1} H^2(\mathbb{D})\otimes_A \mathbb{C}_z \rightarrow \mathcal{M}_\theta\otimes_A \mathbb{C}_z \rightarrow 0.$$

Note that for g in $H^2(\mathbb{D})$, and $|z| < 1$, $(\theta\otimes_A 1)(g\otimes_A 1) = (\theta\cdot g)\otimes_A 1 = g\otimes_A \theta\cdot 1 = \theta(z)(g\otimes_A 1)$, so that $\theta\otimes_A 1$ is onto if $\theta(z) \neq 0$ and is the 0 map when $\theta(z) = 0$. Thus, by exactness $\mathcal{M}_\theta\otimes_A \mathbb{C}_z$ is isomorphic to $H^2(\mathbb{D})\otimes_A \mathbb{C}_z$ when $\theta(z) = 0$ and is 0 otherwise. Since $H^2(\mathbb{D})\otimes_A \mathbb{C}_z$ is already known to be isomorphic to $\mathbb{C}_z$ for $|z|< 1$, we see that $\mathcal{M}_\theta \otimes_A \mathbb{C}_z$ is isomorphic to $\mathbb{C}_z$ whenever $\theta(z) = 0$, and is 0 otherwise.

We also have that when $\theta(z) = 0$, $\theta\otimes_A 1$ is not one-to-one. Thus, injectivity of the map θ does not imply injectivity of the map $\theta\otimes_A 1$. In particular, we see that the exact sequences of theorem 5.12 can not be extended any further to the left.

As we have seen in the examples above much information about the Hilbert module $\mathcal{M}$ can be obtained by forming $\mathcal{M}\otimes_A \mathbb{C}_x$. We shall call this module the <u>localization of $\mathcal{M}$ to x</u>, and denote it by $\mathcal{M}_x$. Of course, it is not just the modules $\{\mathcal{M}_x: x \in \mathcal{M}_A\}$ that are important but how they fit together as x varies. Since we may identify $\mathcal{M}_x$ with $[A_x\cdot\mathcal{M}]^\perp$, which is a subspace of $\mathcal{M}$, any topology on the subspaces of $\mathcal{M}$ endows the set of local modules with a topology, and we may regard the local modules as a <u>sheaf</u> over M_A. We shall use the norm topology on $\mathcal{M}$ to topologize this set and let $\underline{\mathcal{M}}$ denote this sheaf of local modules. Actually, for natural function algebras of holomorphic functions, $\underline{\mathcal{M}}$ has more structure, it is a holomorphic sheaf (cf.[39]).

<u>Example 5.16</u>. Let Ω be an analytic Cauchy domain in $\mathbb{C}$ and fix a Hilbert space $\mathcal{H}$. M.J. Cowen and the first author [38] have defined $\mathbb{B}_n(\Omega)$ to be the set of those operators T on $\mathcal{H}$ such that:

a) Ω is contained in the spectrum of T,

b) $\operatorname{ran}(T-w) = \mathcal{H}$ for w in Ω,

c) the span of $\{\ker(T-w): w \in \Omega\}$ is dense in $\mathcal{H}$,

d) $\dim \ker(T-w) = n$ for all w in Ω.

Typical examples of such operators are the adjoint of the unilateral shift, which belongs to $\mathbb{B}_1(\mathbb{D})$, or the adjoints of the pure subnormal operators of multiplication by z on $H^2_E(\Omega)$ introduced by Abrahamse and the first author which we discussed in Chapter 2. These belong to $\mathbb{B}_n(\overline{\Omega})$, where n is the dimension of the vector bundle E.

If T is in $\mathbb{B}_n(\Omega)$, then the map $w \to \ker(T-w)$ defines a complex vector bundle over Ω. This vector bundle enjoys more structure, namely it is a Hermitian holomorphic vector bundle. These bundles are a complete unitary invariant for these operators. That is, if T and S both belong to $\mathbb{B}_n(\Omega)$, then T and S are unitarily equivalent if and only if their corresponding bundles are equivalent as Hermitian holomorphic vector bundles [38, Theorem 1.14].

As we have observed before, it is the adjoints of operators in familiar classes that belong to $\mathbb{B}_n(\Omega)$ and hence the structure exhibited for a given operator is usually of an anti-holomorphic character and not holomorphic. At the expense of some interchanging of operators for their adjoints and domains for their complex conjugates, we can avoid this for Hilbert modules. The natural algebraic structure for the sheaf $\underline{\mathcal{M}}$ is holomorphy in case A is an algebra of holomorphic functions.

Now let T^* be in $\mathbb{B}_n(\Omega)$ and assume in addition that $\|f(T)\| \leq k\|f\|$ for all f in $R(\Omega)$, so that $\mathcal{H}$ becomes a Hilbert module over $R(\Omega)$ under the action $f \cdot h = f(T)h$.

The maximal ideal space of $R(\Omega)$ is Ω^- and for z in Ω^-, the local module $\mathcal{H}_z = \ker(T^* - \overline{z})$, has dimension n for z in Ω and can be shown to be 0-dimensional for z in $\partial\Omega$.

Thus, the sheaf of local modules restricted to Ω is actually a Hermitian holomorphic vector bundle over Ω, whose equivalence class as a Hermitian holomorphic bundle, determines the unitary equivalence class of the Hilbert module.

If we regard $\mathcal{M}_x = [A_x \cdot \mathcal{M}]^{\perp}$ and assume, in addition, that the span of $\{\mathcal{M}_x : x \in M_A\}$ is dense in $\mathcal{M}$, then each vector h in $\mathcal{M}$ determines a section of this bundle, $x \to h \otimes_A 1_x$, where we use 1_x to denote 1 in $\mathbb{C}_x$. We set $h_x = h \otimes_A 1_x$, and note that our additional hypothesis insures that the map $h \to h_x$ defines a one-to-one map of $\mathcal{M}$ into this space of sections of $\underline{\mathcal{M}}$, and hence the norm on $\mathcal{M}$ endows this space of sections with a norm. Observe that these sections are algebraic sections, in general, and holomorphic when A is an algebra of holomorphic functions. In fact, the holomorphic structure for $\underline{\mathcal{M}}$ is precisely that which makes these sections holomorphic.

Thus, we see that the Hilbert module $\mathcal{M}$ has a model as a space of sections of a sheaf of fairly elementary Hilbert modules above the points in M_A. The module action by f in A on one of these sections is just pointwise multiplication by f(x).

Such structures are perhaps too general to yield much information but it should be possible to identify subclasses of these where one can do business. The results of Cowen and the first author [38], when looked at from this viewpoint, analyze the structure of these objects when A is the algebra $R(\Omega^-)$ and the bundles are assumed to be Hermitian, holomorphic vector bundles.

<u>Definition</u> <u>5.17</u>. Let $\mathcal{M}$ be a Hilbert module for the function algebra A. We say that $\mathcal{M}$ is <u>locally</u> <u>free</u> if $\dim(\mathcal{M}_x) = 0$ for x in ∂A and if the sheaf $\underline{\mathcal{M}}$ of local modules is a finite dimensional vector bundle when restricted to $M_A \backslash \partial A$. Moreover, we assume that the span of $\{\mathcal{M}_x : x \in M_A \backslash \partial A\}$ is dense in $\mathcal{M}$.

<u>Example</u> <u>5.18</u>. Consider $H^2(\mathbb{D}^2)$ as a Hilbert module over $A(\mathbb{D}^2)$ and the submodule, $H^2_{(0,0)}(\mathbb{D}^2) = \{g \in H^2(\mathbb{D}^2) \mid g(0,0) = 0\}$. It is not difficult to show that $H^2(\mathbb{D}^2)$ is locally free. In fact, $H^2(\mathbb{D}^2) \otimes_A \mathbb{C}_x$ is 1-dimensional for every $x = (x_1, x_2)$ in $\mathbb{D}^2$ and the function $k_x(z_1,z_2) = (1-\overline{x}_1 z_1)^{-1} \; (1-\overline{x}_2 z_2)^{-1}$ spans $[A(\mathbb{D}^2)_x \cdot H^2(\mathbb{D}^2)]^\perp$.

Thus, the map $x \to k_x \otimes_A 1$ defines a non-vanishing section of this bundle.

On the other hand, $H^2_{(0,0)}(\mathbb{D}^2) \otimes_A \mathbb{C}_x$ is 1-dimensional for

$x \neq (0,0)$ and 2-dimensional for $x = (0,0)$ and thus is not locally free. If we let P denote orthogonal projection onto $H^2_{(0,0)}(\mathbb{D}^2)$ then $Pk_x = k_x - 1$ spans $[A(\mathbb{D}^2)_x \cdot H^2_{(0,0)}(\mathbb{D}^2)]^\perp$ for $x \neq (0,0)$ while $[A(\mathbb{D}^2)_{(0,0)} \cdot H^2_{(0,0)}(\mathbb{D}^2)]^\perp$ is spanned by the two coordinate functions z_1 and z_2.

Thus, the example indicates that the next stage might be to consider holomorphic sheaves defined by bundles with singularities. We will have more to say about this in the next chapter. We conclude this chapter with an analogue of one final topic from homological algebra.

Definition 5.19. Let $\mathcal{L}$ be a bounded Hilbert module for A. If for every exact sequence of bounded Hilbert modules for A,

$$0 \to \mathcal{M}_2 \xrightarrow{\Phi_1} \mathcal{M}_1 \xrightarrow{\Phi_0} \mathcal{M}_0 \to 0, \text{ we have that}$$

$$0 \to \mathcal{M}_2 \otimes_A \mathcal{L} \xrightarrow{\Phi_0 \otimes_A 1} \mathcal{M}_1 \otimes_A \mathcal{L} \xrightarrow{\Phi_0 \otimes_A 1} \mathcal{M}_0 \otimes_A \mathcal{L} \to 0 \text{ is exact}$$

then $\mathcal{L}$ is flat. If the same conclusion holds with the additional hypotheses that $\mathcal{M}_i$ is contractive and Φ_i is partially isometric, then we call $\mathcal{L}$ hypo-flat.

When theorem 5.12 applies this is equivalent to the statement that $\mathcal{N}_2$ maps onto $\mathcal{N}_1$.

Choosing for each z_0 with $|z_0| < 1$ the inner function $\theta(z) = (z - z_0)/(1 - \bar{z}_0 z)$, we see by example 5.15 that $\mathbb{C}_{z_0}$, $|z_0| < 1$ is not a flat $A(\mathbb{D})$-module.

Similar examples show that if Ω is an analytic Cauchy domain, then $\mathbb{C}_z$, z in Ω; is not a flat $R(\Omega^-)$-module. The following result shows that $\mathbb{C}_z$ is hypo-flat when z is in $\partial\Omega$.

Propositon 5.20. Let x be a strong separating point for A. Then $\mathbb{C}_x$ is hypo-flat.

Proof. Let $\mathcal{M}$ be a contractive Hilbert module for A. Since $\mathcal{M}\otimes_A \mathbb{C}_x$ is a quotient of $\mathcal{M}\otimes\mathbb{C}_x = \mathcal{M}$, by the hypo-projectivity of $\mathbb{C}_x$ (Proposition 4.9), each vector h in $\mathcal{M}\otimes_A\mathbb{C}_x$ lifts to a unique vector $\tilde{h}$ in $\mathcal{M}$ with $\|h\| = \|\tilde{h}\|$ and $f \cdot \tilde{h} = f(x)\tilde{h}$. Set $T_f h = f\cdot h$, and assume $\|f\| = f(x)$. Since $\|\overline{f(x)}h\| \leq \|T_f^* h\| \leq \|f\|\|h\|$, we have that $T_f^* h = \overline{f(x)}h$, that is, h is actually a reducing eigenvector for such an f. But since x is a strong separating point, such functions f generate A. Thus, $T_f h = f(x)h$ and $T_f^* h = \overline{f(x)}h$ for all f in A.

Therefore each vector in $\mathcal{M}\otimes_A\mathbb{C}_x$ corresponds to a reducing eigenvector in $\mathcal{M}$.

The converse is also clear. That is, the vectors in $\mathcal{M}\otimes_A\mathbb{C}_x$ exactly correspond to the reducing eigenvectors for T_f with eigenvalue f(x).

Now if $\mathcal{M}_0 \to \mathcal{M}_1 \to \mathcal{M}_2$ is exact at $\mathcal{M}_1$, h in $\mathcal{M}_1$ maps to 0 in $\mathcal{M}_2$, and h_1 is a reducing eigenvector for T_f with eigenvalue f(x), then h has some pre-image h_0 in $\mathcal{M}_0$ and it is easily checked that h_0 is also a reducing eigenvector.

Thus, $\mathcal{M}_0\otimes_A\mathbb{C}_x \to \mathcal{M}_1\otimes_A\mathbb{C}_x \to \mathcal{M}_2\otimes_A\mathbb{C}_x$ is exact at $\mathcal{M}_1\otimes_A\mathbb{C}_x$. Hence every long exact sequence is preserved by tensoring with $\mathbb{C}_x$, and so $\mathbb{C}_x$ is hypo-flat. □

6 Further thoughts

In this book we have reformulated a part of operator theory in the algebraic language of modules. Many concrete examples of Hilbert modules for natural function algebras have been presented along with a systematic introduction of the class of Šilov modules for general function algebras. The possibility of studying arbitrary Hilbert modules using a Šilov resolution, or indeed a resolution by some tractable class of modules, has been discussed but no new applications were presented. Finally, a localization technique based on the module tensor product was exhibited in the preceding chapter, but again with no examples. In this final chapter we remedy these omissions, at least to some extent. We will provide some sample applications showing how to put it all together and discussing what is possible, or seems to be possible, and what needs to be done. The results described here have been obtained partly in collaboration with G. Misra and K. Yan and more complete details will appear later. Since this is basically work in progress and definitive results are not yet in sight, we do not attempt to provide a full account of this work.

The principal goal in the study of Hilbert modules is to obtain useful invariants up to unitary equivalence or similarity for the modules. Ideally one expects to use a resolution by an understood class of modules for calculating, or determining these invariants. That is the approach from homological algebra that we are modeling. The problem we face, however, is that we have, or understand, few invariants for Hilbert modules for anything other than modules for planar algebras. Hence, we must tackle two problems at once; first that of defining module invariants and second, that of calculating them.

Invariants can be divided into two classes, global and local invariants. For modules over C(X), we consider the dimension function to be a local invariant, for it assigns a Hilbert space of the appropriate dimension "above" each point in X, which is the maximal ideal space of C(X). However, the scalar spectral measure, which describes how these spaces are "glued" together is not local in the same sense. The most basic global invariant, the module spectrum, is just the coordinate-free analogue of the usual notion of joint spectrum. Although the definition could be based on any notion of joint spectrum, we choose to work with that given by Taylor [102].

Let $\mathcal{M}$ be a Hilbert module over the function algebra A. If A is finitely generated as an algebra with generators $\underline{\varphi} = \{\varphi_1, \ldots, \varphi_n\}$, then we define $\mathrm{Spec}_A(\mathcal{M})$ to be the compact subset of the maximal ideal space M_A which is the Taylor spectrum of the n-tuple of commuting operators $\underline{\varphi}$ defined on $\mathcal{M}$ by the module action. More precisely, the generators $\underline{\varphi}$ acting as coordinate functions determine a homeomorphism of M_A with a compact subset of $\mathbb{C}^n$, $\underline{\varphi}(M_A)$. The Taylor spectrum $\sigma(\underline{\varphi})$, which is defined using the Koszul complex, is also a subset of $\mathbb{C}^n$ and can be easily seen to be a subset of $\underline{\varphi}(M_A)$. We define $\mathrm{Spec}_A(\mathcal{M})$ to be the subset $\underline{\varphi}^{-1}(\sigma(\underline{\varphi}))$ of M_A. That this subset is independent of the set of generators chosen follows from basic mapping properties of the Taylor spectrum.

If A is not finitely-generated, then let $\{A_\alpha\}_\alpha$ denote the collection of finitely-generated subalgebras of A partially-ordered by inclusion. Since we have that

$$M_A = \lim_{\overleftarrow{\alpha}} M_{A_\alpha},$$

then we see that

$$\mathrm{Spec}_A(\mathcal{M}) = \lim_{\overleftarrow{\alpha}} \mathrm{Spec}_{A_\alpha}(\mathcal{M})$$

can be identified as a subset of M_A. Again properties of the

Taylor spectrum show that this all makes sense and we summarize this in the following statement.

Proposition 6.1 Let $\mathcal{M}$ be a Hilbert module over the function algebra A. Then the module spectrum $\mathrm{Spec}_A(\mathcal{M})$ is well-defined and is a similarity invariant for $\mathcal{M}$.

Before continuing let us raise a natural question about the module spectrum. A compact subset X of M_A determines the algebra $\mathrm{Rat}_A(X)$ of rational functions in A with poles off X. That is, $\mathrm{Rat}_A(X)$ consists of the continuous functions on X that can be written in the form ψ/φ, where φ and ψ are in A and φ doesn't vanish on X. Now a Hilbert module $\mathcal{M}$ for A can be extended to a module for the algebra $\mathrm{Rat}_A(\mathrm{Spec}_A(\mathcal{M}))$ in a straight-forward manner. The question is whether it extends to the closed function algebra, or equivalently, whether it is bounded. If $\mathcal{M}$ is contractive for A, then $\mathcal{M}$ need not be contractive for $\mathrm{Rat}_A(\mathrm{Spec}_A(\mathcal{M}))$, or even bounded. When A is singly-generated by φ, this is just the statement that the spectrum of the operator φ, need not be a K-spectral set for φ, for any K. However, if A = C(X), $\mathcal{M}$ always is a bounded module for $\mathrm{Rat}_A(\mathrm{Spec}_A(\mathcal{M}))$.

Problem 6.2. For which algebras A, is $\mathcal{M}$ always a bounded module for R = $\mathrm{Rat}_A(\mathrm{Spec}_A(\mathcal{M}))$? If so, then how does the constant C_A depend on A, where

$$C_A = \sup_{\mathcal{M}} \frac{\|\mathcal{M}\|_R}{\|\mathcal{M}\|_A} \ ?$$

The extension problem for modules can be important in case the classification problem for modules for the super algebra is much simpler than that for the original algebra. There is one case of this that warrants a definition. A contractive Hilbert module $\mathcal{M}$ for A is said to be reductive if the operators on $\mathcal{M}$ defined by the module action are all normal. In that case, $\mathcal{M}$

extends to a <u>contractive</u> module for $C(\mathrm{Spec}_A(\mathcal{M}))$ and the analysis of Chapter 1 then applies.

Another global invariant for a Hilbert module $\mathcal{M}$ is its essential spectrum $\mathrm{Spec}^e_A(\mathcal{M})$ which is defined in the same manner as $\mathrm{Spec}_A(\mathcal{M})$ but using the image of the multiplication operators in the Calkin algebra $\mathcal{Q}(\mathcal{M})$ instead of in $\mathcal{L}(\mathcal{M})$. In analogy with the notion of reductive module, we can define an essentially reductive module. The Hilbert module $\mathcal{M}$ for A is said to be <u>essentially reductive</u> if the operators defined on $\mathcal{M}$ by the module action are all essentially normal. For an essentially reductive module $\mathcal{M}$ we obtain a $*$-homormorphism

$$\tau_{\mathcal{M}}: \quad C(\mathrm{Spec}^e_A(\mathcal{M})) \to \mathcal{Q}(\mathcal{M})$$

which determines an element $[\mathcal{M}] = [\tau_{\mathcal{M}}]$ in the K-homology group $K_1(\mathrm{Spec}^e_A(\mathcal{M}))$. The element $[\mathcal{M}]$ can be viewed as a primordal characteristic class for $\mathcal{M}$ and is a global unitary invariant for $\mathcal{M}$. (In fact, $[\mathcal{M}]$ is a deformation invariant for $\mathcal{M}$.)

Do essentially reductive Hilbert modules exist? The result of Berger-Shaw [24] can be reformulated to provide an affirmative answer for the planar algebras $R(\Omega)$.

<u>Theorem 6.3 (Berger-Shaw)</u> Let Ω be a bounded domain in $\mathbb{C}$ and let $\mathcal{M}$ be a subnormal Hilbert module over $R(\Omega)$ of finite rank. Then $\mathcal{M}$ is essentially reductive.

The analogue of this result for non-planar algebras is false without further hypotheses. Perhaps somewhat surprisingly, an appropriate condition involves the smoothness of the topological boundary of $\mathrm{Spec}_A(M)$.

<u>Problem 6.4.</u> Let Ω be a bounded domain in $\mathbb{C}^n$ and let $\mathcal{M}$ be a subnormal Hilbert module over $R(\Omega)$ of finite rank. Does it follow from the assumption that $\partial\, \mathrm{Spec}_{R(\Omega)}(\mathcal{M})$ is smooth that $\mathcal{M}$ is essentially reductive?

This problem is related to work of Curto-Muhly [43] and recent work of Salinas, Sheu and Upmeier.

Along different lines, K. Yan and the first author have shown recently that the Berger-Shaw theorem can be extended to some Hilbert modules for which the module spectrum lies on an algebraic curve. In this case the critical dimension is not that of the function algebra but rather that of the module.

There is another very different setting in which one can conclude that a Hilbert module $\mathcal{M}$ is essentially reductive and even calculate $[\mathcal{M}]$. Consider the Hardy module $H^2(\mathbb{D}^2)$ for the bidisk algebra $A(\mathbb{D}^2)$ and the submodule $\mathcal{N}_{j,k} = \{f \in H^2(\mathbb{D}^2): f(z^j,z^k) \equiv 0 \text{ for } z \in \mathbb{D}\}$ for positive integers j and k. Let $Q_{j,k}$ denote the quotient module $H^2(\mathbb{D}^2)/\mathcal{N}_{j,k}$. In [56] Misra and the first author showed that $Q_{j,k}$ is essentially reductive. This was greatly generalized by Clark in [36] to quotient modules over the polydisk algebra $A(\mathbb{R})$, where $\mathcal{N}_{j,k}$ is replaced by the submodule of functions f in $H^2(\mathbb{D}^n)$ which satisfy

$$f(z_1,\ldots,B_i(z),\ldots;B_j(z),\ldots z_n) \equiv 0 \text{ for } z \text{ in } \mathbb{D}$$

and $$z_1,\ldots,z_{i-1}, z_{i+1},\ldots,z_{j-1}, z_{j+1},\ldots,z_n \text{ in } \mathbb{D}$$

for all integers satisfying $1 \leq i < j \leq n$, where $B_1,B_2,\ldots,B_n$ are finite Blaschke products.

Clark also calculates the characteristic class $[Q_{j,n}]$ for the quotient module although he does not state the result in this language. We should add that neither result is the main thrust of his paper which we will comment on later in this chapter.

We can take the Chern character of $[\mathcal{M}]$ in $K_1(\mathrm{Spec}_A^e(\mathcal{M}))$ to obtain an ordinary homology class which is also a global unitary invariant and which is closely related to the Fredholm index of the operators defined by the module action. An important part of the work of Helton-Howe [73] and of Pincus-Carey [31] can be viewed as the analogue of Chern-Weil theory for these characteristic classes based on cyclic cohomology which is defined using trace forms. This topic is especially interesting since the latter objects also have a local character. A penetrating explanation should be possible in the language of Hilbert modules and we will have more to say about this at another time.

We now turn to more basic and elementary local invariants. Local invariants are those that are local relative to the maximal (or prime) ideals of A, or what is equivalent, to the ideals in $\mathrm{Spec}_A(\mathcal{M})$, since $\mathcal{M}$ is locally trivial outside $\mathrm{Spec}_A(\mathcal{M})$. Basically one is trying to define the proper notion of the multiplicity of $\mathcal{M}$ at the ideal in $\mathrm{Spec}_A(\mathcal{M})$. However, as we saw in Chapter 1, this is not always simple. For example, in the case of C(X) the multiplicity must also involve the scalar spectral measure on M_A. For natural function algebras of holomorphic functions, however, the appropriate notion of multiplicity at an interior point of M_A is algebraic in character for similarity, but (finite-dimensional) Hilbert space geometric for unitary equivalence. Although this is not the complete story for the invariants of a Hilbert module, we concentrate on that here, leaving boundary behavior and non-holomorphic phenomena to future studies. We begin with a useful definition.

A Hilbert module $\mathcal{L}$ for the function algebra A is said to be a <u>local module for x_0</u> if $\mathcal{L}$ is finite dimensional and $\mathrm{Spec}_A(\mathcal{L})$ consists of the point x_0.

The simplest example of a local module for the point x_0 is, of course, the module $\mathbb{C}_{x_0}$. Although we focus in what follows mainly on such local modules, more general examples are needed for a comprehensive treatment. For a natural function algebra of holomorphic functions on some domain in $\mathbb{C}^n$, the local modules for an interior point are often the same as the local modules for $\mathbb{C}[z_1,\ldots,z_n]$. And although classifying the local modules for $\mathbb{C}[z_1,\ldots,z_n]$ is not trivial nor, perhaps, even well-understood, it is an algebraic problem and hence we view it as tractable.

Localization of a Hilbert module $\mathcal{M}$ to a point x_0 can be accomplished by considering the module tensor product $\mathcal{M}\otimes_A\mathcal{L}$, where $\mathcal{L}$ is a local module for x_0. If $\mathcal{M}$ has finite rank, then $\mathcal{M}\otimes_A\mathcal{L}$ is also a local module for x_0, probably not isomorphic to $\mathcal{L}$, and the invariants which determine $\mathcal{M}\otimes_A\mathcal{L}$ are then invariants for $\mathcal{M}$. For similarity one uses algebraic invariants, while for unitary equivalence one uses Hilbert space geometric invariants. We illustrate this with an interesting application extending the result of Agrawal, Clark, and Douglas stated in Chapter 2.

Recall that there is a one-to-one correspondence between submodules $\mathcal{M}$ of $H^2(\mathbb{D}^2)$ of finite codimension and ideals I in $\mathbb{C}[z_1,\ldots z_n]$ of finite codimension for which the zero variety $Z(I) = \{\underline{z} \in \mathbb{C}^n\colon p(\underline{z}) = 0 \text{ for } p \in I\}$ is assumed to be a subset of the open polydisk $\mathbb{D}^n$. In such a case one has $I = \mathcal{M} \cap \mathbb{C}[z_1,\ldots z_n]$.

In [8] it was shown that two such submodules $\mathcal{M}_1$ and $\mathcal{M}_2$ are unitarily equivalent if and only if they coincide. Using localization techniques and a result from algebraic geometry we have been able to extend this to the following.

<u>Theorem 6.5.</u> If $\mathcal{M}_1$ and $\mathcal{M}_2$ are two submodules of finite codimension in $H^2(\mathbb{D}^n)$ $(n>1)$, then $\mathcal{M}_1$ and $\mathcal{M}_2$ are quasi-similar if and only if they are equal.

We only sketch the proof, since shortly we shall be proving something more general. The proof proceeds by observing that the corresponding ideals $I_i = \mathcal{M}_i \cap \mathbb{C}[z_1,\ldots z_n]$

for $i = 1,2$ are "algebraic models" for $\mathcal{M}_i$ via localization. More precisely, let us view the local module $\mathcal{L}$ for $\underline{z}$ in $\mathbb{D}^n$ as a local module for both $A(\mathbb{D}^n)$ and $\mathbb{C}[z_1,\ldots z_n]$. Then it can be shown that the two modules $\mathcal{M}_i \otimes_{A(\mathbb{D}^n)} \mathcal{L}$ and $I_i \otimes_{\mathbb{C}[z_1,\ldots z_n]} \mathcal{L}$ are isomorphic for $i = 1,2$.

Now $\mathcal{M}_1$ and $\mathcal{M}_2$ are quasi-similar if there exist bounded module maps $X\colon \mathcal{M}_1 \to \mathcal{M}_2$ and $Y\colon \mathcal{M}_2 \to \mathcal{M}_1$ such that both XY and YX have dense range and no null space. Due to the finite dimensionality of $\mathcal{M}_1 \otimes_{A(\mathbb{D}^n)} \mathcal{L}$ and $\mathcal{M}_2 \otimes_{A(\mathbb{D}^n)} \mathcal{L}$ as vector spaces, it follows that $X \otimes_{A(\mathbb{D}^n)} 1_{\mathcal{L}}$ is a module isomorphism by applying Theorem 5.12. Hence the modules $I_1 \otimes_{\mathbb{C}[z_1,\ldots z_n]} \mathcal{L}$ and $I_2 \otimes_{\mathbb{C}[z_1,\ldots z_n]} \mathcal{L}$ are isomorphic for every local module $\mathcal{L}$ for any point $\underline{z}$ in $\mathbb{D}^n$. However, since $Z(I_1)$ and $Z(I_2)$ are contained in $\mathbb{D}^n$, it follows that

$I_1 \otimes_{\mathbb{C}[z_1,\ldots z_n]} \mathcal{L} \cong \mathbb{C}[z_1,\ldots z_n] \otimes_{\mathbb{C}[z_1,\ldots z_n]} \mathcal{L} \cong I_2 \otimes_{\mathbb{C}[z_1,\ldots z_n]} \mathcal{L}$ for a local module $\mathcal{L}$ for $\mathbb{C}[z_1,\ldots z_n]$ for a point $\underline{z}$ outside $\mathbb{D}^n$. Combining the statements we see that I_1 and I_2 are locally isomorphic.

The result now follows by appealing to a result from algebraic geometry based on work of M. Artin and Grothendieck and communicated to us by H. Sah. This result states that two ideals I_1 and I_2 in $\mathbb{C}[z_1,\ldots z_n]$ which satisfy

$$I_1 \otimes_{\mathbb{C}[z_1,\ldots z_n]} \mathcal{L} \cong I_2 \otimes_{\mathbb{C}[z_1,\ldots z_n]} \mathcal{L}$$

for every finite dimensional module $\mathcal{L}$ over $\mathbb{C}[z_1,\ldots z_n]$ must be equal if the zero sets $Z(I_1)$ and $Z(I_2)$ have codimension greater than one in $\mathbb{C}^n$.

The beauty of the preceding proof is that it will extend to more general domains in $\mathbb{C}^n$ such as the ball or many holomorphically convex domains. Moreover, the technique can be applied to submodules $\mathcal{M}$ obtained from the closure of ideals I in $\mathbb{C}[z_1,\ldots z_n]$ so long as $Z(I)$ has codimension greater than one in $\mathbb{D}^n$. Thus for $n \geq 3$, there are examples to which the proof technique applies with no finite dimensional hypothesis at all. What the assumption on $Z(I)$ excludes is the principal ideals for which the result is not true.

In order to state a fairly general result that these techniques imply we need to recall several concepts. Given a bounded open set G in $\mathbb{C}^n$ we say that a Hilbert space $\mathcal{H}$ is a <u>reproducing kernel Hilbert space on G</u>, provided that every element of $\mathcal{H}$ is a function that is analytic on G, the evaluation maps $E_\lambda(h) = h(\lambda)$ are bounded linear functionals for each λ in G, and multiplication by each of the coordinate functions is a bounded, linear map.

It is not hard to see that the map $\lambda \rightarrow E_\lambda$ is weak*-analytic and hence norm analytic as a map into the dual space of $\mathcal{H}$. But since $(\frac{\partial}{\partial z_i} E_\lambda)(h) = (\frac{\partial}{\partial z_i} h)(\lambda)$, we see that evaluations of partial derivatives of all orders are bounded linear functionals on $\mathcal{H}$.

The other concepts we need to recall are from commutative algebra. To simplify notation we set $\mathcal{R} = \mathbb{C}[z_1,\ldots,z_n]$. Every ideal I in $\mathcal{R}$ has a <u>primary decomposition</u>, that is, $I = \bigcap_{j=1}^{m} I_j$ where each I_j is a <u>primary</u> ideal with associated prime ideal P_j. While the set $\{I_j\}_{j=1}^m$ is not uniquely detemined by I the set $\{P_j\}_{j=1}^m$ is uniquely determined by I and these are called the <u>associated prime ideals</u> of I, [13, Chapter 4]. We let $Z(I) = \{\lambda \in \mathbb{C}^n : p(\lambda) = 0$ for every p in $I\}$, and recall that prime ideals have the property that $P = \{p \in \mathcal{R} : p(\lambda) = 0$ for all $\lambda \in Z(P)\}$. Also, recall that if S is a subset of $\mathcal{R}$ which is multiplicatively closed, then [13, Chapter 3] we may form a quotient ring $S^{-1}(\mathcal{R}) = \{p/q : p \in \mathcal{R}, q \in S\}$. If $\lambda \in \mathbb{C}^n$ and $\mathcal{M}_\lambda = \{p \in \mathcal{R} : p(\lambda) = 0\}$ denotes the maximal ideal of polynomials that vanish at λ, then $S_\lambda = \mathcal{R} \backslash \mathcal{M}_\lambda$ is a multiplicatively closed set.

Finally, if $\lambda \in G$, and $\alpha = (\alpha_1,\ldots,\alpha_n)$ is an n-tuple of non-negative integers, and h is analytic, we set

$$D_\lambda^\alpha h = \frac{\partial^{\alpha_1}}{\partial z_1^{\alpha_1}} \cdot \ . \ . \ \frac{\partial^{\alpha_n}}{\partial z_n^{\alpha_n}} h(\lambda).$$

and set $|\alpha| = \alpha_1+\ldots+\alpha_n$.

<u>Theorem 6.6.</u> Let $\mathcal{H}$ be a reproducing kernel Hilbert space on G, and let I be an ideal in $\mathcal{R}$ with associated prime ideals $\{P_i\}_{i=1}^m$. If $Z(P_i) \cap G \neq 0$ for all $i=1,\ldots,m$, then $[I]\cap\mathcal{R} = I$, where [I] denotes the closure of I in $\mathcal{H}$.

<u>Proof.</u> Let $J = [I] \cap\mathcal{R}$, then clearly J is an ideal in $\mathcal{R}$ that contains I.

Fix λ in G and a non-negative integer j. Since $\{D_\lambda^\alpha: |x| \leq j\}$ is a finite family of continuous linear functionals, for any p in J there exists q in I such that $D_\lambda^\alpha p = D_\lambda^\alpha q$, for $|\alpha| \leq j$, which implies that (p-q) is in $\mathcal{M}_\lambda^j \cap J$, where $\mathcal{M}_\lambda^j$ is the product of the maximal ideal $\mathcal{M}_\lambda$ taken with itself j times. Thus, $J \subseteq I + \mathcal{M}_\lambda^j \cap J$ for all j.

By the Artin-Rees Lemma [13,Corollary 10.10], there is an integer k, such that for $j \geq k$ $\mathcal{M}_\lambda^j \cap J = \mathcal{M}^{j-k}{}_\lambda \cdot (\mathcal{M}_\lambda^k \cap J)$. Setting $j = k+1$, we have that $J \subseteq I + \mathcal{M}_\lambda \cdot J$. Passing to the quotient ring $S_\lambda^{-1}(\mathcal{R})$, we see that, $S_\lambda^{-1}(J) \subseteq S_\lambda^{-1}(I) + S_\lambda^{-1}(\mathcal{M}_\lambda) \cdot S_\lambda^{-1}(J)$. Since $S_\lambda^{-1}(\mathcal{M}_\lambda)$ is the unique maximal ideal in $S_\lambda^{-1}(\mathcal{R})$, we may apply Nakayama's Lemma [13, Corollary 2.7] to deduce that $S_\lambda^{-1}(J) \subseteq S_\lambda^{-1}(I)$ for any λ in G. Thus,

$$J \subseteq \cap \{S_\lambda^{-1}(I) \cap \mathcal{R}: \lambda \in G\}.$$

However, for $\lambda \in Z(P_j) \cap G$, we have that $S_\lambda^{-1}(I) \cap \mathcal{R} \subseteq S_\lambda^{-1}(I_j) \cap \mathcal{R} = I_j$, by [13, Proposition 4.9], and so we see that

$$\{S_\lambda^{-1}(I)\cap R: \lambda\varepsilon G\} = I.$$

Hence $J = I$, which is what we wished to prove. □

If I is an ideal of finite-codimension, then the associated prime ideals are the maximal ideals corresponding to the points in $Z(I)$. Thus, the above theorem generalizes the second half of the theorem of Ahern-Clark (Theorem 2.23).

We let $\mathcal{C}$ denote the family of ideals I in $\mathcal{R}$ such that $[I] \cap \mathcal{R} = I$, where $[I]$ denotes the closure of I in $\mathcal{H}$. By the above theorem any ideal satisfying the above zero set condition is in $\mathcal{C}$.

<u>Lemma 6.7.</u> Let I_1 and I_2 be ideals in $\mathcal{R}$.
Then:

i) if I_1, I_2 are in $\mathcal{C}$, then $I_1 \cap I_2$ is in $\mathcal{C}$,

ii) if I_1 is in $\mathcal{C}$ and $I_1 \subseteq I_2$ with $\dim (I_2/I_1)$ finite, then I_2 is in $\mathcal{C}$,

iii) if I_1, I_2 are in $\mathcal{C}$, $\dim(\mathcal{R}/I_1)$ is finite and $Z(I_1) \subseteq G$, then $I_1 \cdot I_2$ is in $\mathcal{C}$.

<u>Proof.</u> We have that

$$I_1 \cap I_2 \subseteq [I_1 \cap I_2] \cap \mathcal{R} \subseteq ([I_1]\cap\mathcal{R}) \cap ([I_2] \cap \mathcal{R}) \subseteq I_1 \cap I_2,$$

which proves i).

To prove ii), write $I_2 = I_1 + \mathcal{F}$ with $\dim(\mathcal{F})$ finite. Then $[I_2] = [I_1] + \mathcal{F}$, and so,

$$[I_2] \cap \mathcal{R} \subseteq ([I_1] + \mathcal{F}) \cap \mathcal{R} \subseteq [I_1] \cap \mathcal{R} + \mathcal{F} \subseteq I_1 + \mathcal{F} \subseteq I_2,$$

which proves ii).

To prove iii), note that by the above theorem, I_1^k is in $\mathcal{C}$ for all k. Applying the Artin-Rees lemma again, we have that for k sufficiently large $I_1^{k+1} \cap I_2 = I_1 \cdot (I^k \cap I_2) \subseteq I_1 \cdot I_2$. Since $\dim (\mathcal{R}/I_1^{k+1})$ is finite and I_2 is finitely generated, $\dim (I_2/I_1^{k+1} \cdot I_2)$ is finite. But since,
$I_1^{k+1} \cdot I_2 \subseteq I_1^{k+1} \cap I_2 \subseteq I_1 \cdot I_2 \subseteq I_2$, we have that $\dim(I_1 \cdot I_2 / I_1^{k+1} \cap I_2)$ is finite. Since $I_1^{k+1} \cap I_2$ is in $\mathcal{C}$ by i), by ii) we have that $I_1 \cdot I_2$ is in $\mathcal{C}$. □

Let $A(G^-)$ be the closure of $\mathcal{R}$ in the continuous functions on G^-, $C(G^-)$.

__Theorem 6.8.__ Let $\mathcal{H}$ be a reproducing kernel Hilbert space on G, and assume that $\mathcal{H}$ is also a bounded, Hilbert $A(G^-)$-module. If I is in $\mathcal{C}$, and $\mathcal{L}$ is a finite dimensional Hilbert $A(G^-)$-module, with $\text{Spec}_{A(G^-)}(\mathcal{L})$ contained in G, then $[I]\otimes_{A(G^-)}\mathcal{L}$ and $I\otimes_{\mathcal{R}}\mathcal{L}$ are similar as $\mathcal{R}$-modules.

__Proof.__ We may assume that $\mathcal{L}$ has a cyclic vector e_0, and let $J_1 \subseteq A(G^-)$ be the annihilating ideal. If we let $J = J_1 \cap \mathcal{R}$, then, using the fact that J is dense in J_1, $\mathcal{R}/J$ is isomorphic to $\mathcal{L}$ as $\mathcal{R}$-modules. It is easily checked that $\text{Spec}_{A(G^-)}(\mathcal{L}) = Z(J)$.

By theorem 5.14, $[I] \otimes_{A(G^-)} \mathcal{L}$ and $[I]/[J_1 \cdot I]$ are similar as $A(G^-)$-modules. Since this space is finite dimensional, the quotient map $I \to [I]/[J_1 \cdot I]$ is onto with kernel $[J_1 \cdot I] \cap \mathcal{R}$. Since $[J_1 \cdot I] = [J \cdot I]$, $Z(J) \subseteq G$ and $\dim(\mathcal{R}/J)$ finite by the above lemma, $[J \cdot I] \cap \mathcal{R} = J \cdot I$. Hence, $[I]/[J\cdot I]^-$ and $I/(J \cdot I)$ are similar $\mathcal{R}$-modules.

Finally, from algebra (or duplicating the proof of theorem 5.14), we know that $I/(J \cdot I)$ and $I \otimes_{\mathcal{R}} (\mathcal{R}/J)$ are similar as $\mathcal{R}$-modules. Thus, $I \otimes_{\mathcal{R}}\mathcal{L}$ and $I \otimes_{A(G^-)} \mathcal{L}$ are similar. □

We have now done all the preliminary analysis necessary to prove our principal theorem. Before proceeding we formally state the result from algebraic geometry which we shall need, see [67].

__Theorem 6.9.__ Let $\mathcal{A}$ be a regular Noetherian ring, containing a field k with $\dim_k(\mathcal{M}/\mathcal{M}^2)$ finite for each maximal ideal $\mathcal{M}$, and let I and J be ideals in $\mathcal{A}$ so that their associated prime ideals all have height at least 2. If $I\otimes_{\mathcal{A}}\mathcal{L}$ is similar to $J \otimes_{\mathcal{A}} \mathcal{L}$ for every finite dimensional $\mathcal{A}$-module, then $I = J$.

In the cases in which we are interested the condition on the heights is equivalent to the requirement that Z(I) and Z(J) have codimension at least 2 in $\mathbb{C}^n$.

We need one last preliminary algebraic lemma. We assume the reader is familiar with [13, Chapter 3].

A set G is <u>rationally</u> <u>convex</u> if for every λ outside G there exists a polynomial p with $p(\lambda) = 0$, but $p(z) \neq 0$ for all z in G.

<u>Lemma</u> <u>6.10</u>. Let G be a bounded, open rationally convex set in $\mathbb{C}^n$, and let $S = \{p \in \mathcal{R}: p(\lambda) \neq 0 \text{ for all } \lambda \text{ in } G\}$. Then the only maximal ideals in $S^{-1}(\mathcal{R})$ are those of the form $S^{-1}(\mathcal{M}_\lambda) = \{\frac{p}{q} : q \in S,\ p(\lambda) = 0\}$ for λ in G. Consequently, the only finite dimensional cyclic $S^{-1}(\mathcal{R})$-modules are of the form $S^{-1}(\mathcal{R}/J) = S^{-1}(\mathcal{R})/S^{-1}(J)$ for J an ideal in $\mathcal{R}$ with $Z(J) \subseteq G$.

<u>Proof</u>. It is standard that the only maximal ideals in $S^{-1}(\mathcal{R})$ are those of the form $S^{-1}(\mathcal{M}_\lambda)$ for $\mathcal{M}_\lambda$ a maximal ideal in $\mathcal{R}$ with $\mathcal{M}_\lambda \cap S = 0$ [13, Proposition 3.11 iv)]. Clearly, if λ is in G, then $\mathcal{M}_\lambda \cap S = 0$ so $S^{-1}(\mathcal{M}_\lambda)$ is a maximal ideal in $\mathcal{R}$. Conversely, if λ is not in G, then there exists a polynomial p with $p(\lambda) = 0$ but $p(z) \neq 0$ for all z in G. Hence, p is in $\mathcal{M}_\lambda \cap S$, and so $S^{-1}(\mathcal{M}_\lambda)$ is not a (non-trivial) maximal ideal in $S^{-1}(\mathcal{R})$ when λ is in G.

Finally, if $\mathcal{L}$ is a finite dimensional cyclic $S^{-1}(\mathcal{R})$-module, then $\mathcal{L} = S^{-1}(\mathcal{R})/J_1$ for some ideal J_1 in $S^{-1}(\mathcal{R})$, and $J_1 = S^{-1}(J)$ for $J = J_1 \cap \mathcal{R}$ an ideal in $\mathcal{R}$. It is easily checked if $J \subseteq \mathcal{M}_\lambda$, then $S^{-1}(\mathcal{M}_\lambda)$ is a non-trivial maximal ideal in $S^{-1}(\mathcal{R})$. Hence, λ is in G and so $Z(J) \subseteq G$. □

<u>Theorem 6.11.</u> Let G be a bounded, open rationally convex set in $\mathbb{C}^n$, let $\mathcal{H}$ be a reproducting kernel Hilbert space, which is also a bounded $A(G^-)$-module, and let I_1, I_2 be ideals in $\mathcal{R}$ such that $Z(P) \cap G$ is non-empty for each of their associated

prime ideals P, and such that $Z(I_i)$ has codimension at least 2 in $\mathbb{C}^n$, i = 1, 2. If $[I_1]$ and $[I_2]$ are quasi-similar submodules of $\mathcal{H}$, then $I_1 = I_2$.

Proof. Let $X: [I_1] \to [I_2]$, , $Y: [I_2] \to [I_1]$ be the $A(G^-)$-module maps which implement the quasi-similarity. If $\mathcal{L}$ is a finite dimensional $A(G^-)$-module, then $[I_1] \otimes_{A(G^-)} \mathcal{L}$ and $[I_2] \otimes_{A(G^-)} \mathcal{L}$ are similar, since by Theorem 5.12, $X \otimes_{A(G^-)} 1$ and $Y \otimes_{A(G^-)} 1$ are mutual inverses.

Combining Theorems 6.5 and 6.8, we have that $I_1 \otimes_{\mathcal{R}} \mathcal{L}$ and $I_2 \otimes_{\mathcal{R}} \mathcal{L}$ are similar for every finite dimensional $A(G^-)$-module $\mathcal{L}$, provided that $\mathrm{Spec}_{A(G^-)}(\mathcal{L})$ is contained in G. Consequently, $S^{-1}(I_1 \otimes_{\mathcal{R}} \mathcal{L}) = S^{-1}(I_1) \otimes_{S^{-1}(\mathcal{R})} S^{-1}(\mathcal{L})$ is similar to $S^{-1}(I_2) \otimes_{S^{-1}(\mathcal{R})} S^{-1}(\mathcal{L})$ as $S^{-1}(\mathcal{R})$-modules, where S is as in lemma 6.10. By lemma 6.10, every finite dimensional $S^{-1}(\mathcal{R})$-module is of the form $S^{-1}(\mathcal{L})$ for some $\mathcal{L}$ as above.

Hence, we may apply Theorem 6.10 with $\mathcal{A} = S^{-1}(\mathcal{R})$ to conclude that $S^{-1}(I_1) = S^{-1}(I_2)$. But by [13, Proposition 4.9] the zero set condition on I_1 and I_2 guarantees that

$$I_1 = S^{-1}(I_1) \cap \mathcal{R} = S^{-1}(I_2) \cap \mathcal{R} = I_2. \qquad \square$$

The preceding result makes clear the importance of the ideals in $\mathbb{C}[z_1,\dots z_n]$ to the study of Hilbert modules over natural function algebras. As some reflection will reveal, the structure of such ideals is not nearly so simple in case $N > 1$ when $\mathbb{C}[z_1,\dots z_n]$ is not a principal ideal domain. However, there is a substantial body of literature in commutative algebra and algebraic geometry directed to this. We will not continue further in this direction now but believe it will be important in future developments.

Before continuing we point out that the previous results can be interpreted in terms of the sheaf $\underline{\mathcal{M}}$ of local modules.

In case $\mathcal{M}$ is the closure of an ideal I in $\mathbb{C}[z_1,\ldots z_n]$ of finite codimension, then $\mathcal{M}$ is a holomorphic bundle over $\mathbb{D}^n\backslash Z(I)$. The sheaf $\mathcal{M}$ has singularities at the points of $Z(I)$ which, as was shown above, characterize I completely.

In the case of planar algebras, most notably the disk algebra, there is a successful approach to the study of contractive Hilbert modules via the canonical model theory of Sz-Nagy and Foiaş. We want to show that this can be obtained in a natural manner via the study of Šilov resolutions. We do not try to handle the most general case.

For finite dimensional Hilbert spaces $\mathcal{E}$ and $\mathcal{E}_*$, let $\theta(z)$ be an inner function $\theta(z)\colon \mathcal{E} \to \mathcal{E}_*$ defined for z in $\mathbb{D}$. That is, $\theta(z)$ is a bounded holomorphic $\mathcal{L}(\mathcal{E},\mathcal{E}_n)$-valued function on $\mathbb{D}$ having unitary-valued radial limits on $\mathbb{T} = \partial\mathbb{D}$ almost everywhere. Then we can view θ as defining an isometry

$\theta\colon\ H^2_{\mathcal{E}}(\mathbb{D}) \to H^2_{\mathcal{E}_*}(\mathbb{D})$ that will be an $A(\mathbb{D})$ module map. If we set

$\mathcal{M}_\theta = H^2_{\mathcal{E}_*}(\mathbb{D})/\theta\, H^2_{\mathcal{E}}(\mathbb{D})$, then $\mathcal{M}_\theta$ is a contractive quotient module

for $A(\mathbb{D})$ and we have the Šilov resolution

$$0 \leftarrow \mathcal{M}_\theta \leftarrow H^2_{\mathcal{E}_*} \leftarrow H^2_{\mathcal{E}} \leftarrow 0$$

where the left map is the quotient map and the right is defined as multiplication by θ. We will denote the latter module map by X.

The basic problem is how to study $\mathcal{M}_\theta$ using only the last part of the resolution $H^2_{\mathcal{E}_*} \to H^2_{\mathcal{E}}$. Now we, of course, know the answer we are seeking, but we want to obtain it in a systematic module theoretic manner. We proceed using localization as follows.

For each w in $\mathbb{D}$ we can tensor with $\mathbb{C}_w$ to obtain the exact sequence

$$0 \leftarrow \mathcal{M}_\theta \otimes_{A(\mathbb{D})} \mathbb{C}_w \leftarrow H^2_{\mathcal{E}_*} \otimes_{A(\mathbb{D})} \mathbb{C}_w \leftarrow H^2_{\mathcal{E}} \otimes_{A(\mathbb{D})} \mathbb{C}_w .$$

Since $H^2_{\mathcal{E}} \otimes_{A(\mathbb{D})} \mathbb{C}_w \cong \mathbb{C}_w \otimes_{\mathbb{C}} \mathcal{E}$ but not canonically, we have the localized module map

$$\mathcal{E}_* \cong \mathbb{C}_w \otimes_{\mathbb{C}} \mathcal{E}_* \xleftarrow{X \otimes_{A(\mathbb{D})^1} \mathbb{C}_w} \mathbb{C}_w \otimes_{\mathbb{C}} \mathcal{E} \cong \mathcal{E}.$$

Now what invariants are determined by a Hilbert space operator T_w in $\mathcal{L}(\mathcal{E}, \mathcal{E}_*)$ for finite dimensional spaces $\mathcal{E}$ and $\mathcal{E}_*$? Basically there are just the eigenvalues of T^*T and TT^* which in this case, are the same since $\dim \mathcal{E} = \dim \mathcal{E}_*$. However, in this case although the identification of $H^2_{\mathcal{E}} \otimes_{A(\mathbb{D})} \mathbb{C}_w$ with $\mathcal{E}$ is not canonical, we can reduce the ambiguity to a single unitary between $\mathcal{E}$ and $\mathcal{E}_*$ since we can use the same identification of $H^2_{\mathcal{E}}(\mathbb{D}) \otimes_{A(\mathbb{D})} \mathbb{C}_w$ and $H^2_{\mathcal{E}_*}(\mathbb{D}) \otimes_{A(\mathbb{D})} \mathbb{C}_w$ making use of the identity $H^2_{\mathcal{E}}(\mathbb{D}) = H^2(\mathbb{D}) \otimes \mathcal{E}$. Then T_w can be defined uniquely up to a unitary V on $\mathcal{E}$ and one can show that single unitary V can be chosen such that $T_w = \theta(w)$ for all w in $\mathbb{D}$. Thus we have a module theoretic derivation of the characteristic operator function of Sz-Nagy and Foiaş.

But it is not quite as simple as it appears. There is more going on as the following example shows. Let p(z) be a polynomial with zeros in $\mathbb{D}$. Then $pB^2(\mathbb{D})$ is a closed submodule of the Bergman module $B^2(\mathbb{D})$. If we let $\mathcal{N}_p$ denote the quotient module defined by $B^2(\mathbb{D})/pB^2(\mathbb{D})$, then we have the exact sequence

$$0 \leftarrow \mathcal{N}_p \leftarrow B^2(\mathbb{D}) \leftarrow pB^2(\mathbb{D}) \leftarrow 0,$$

where the left module map is the quotient map and the right module map is the inclusion which we will denote by Y. In analogy with what we just did, we localize using $\mathbb{C}_w$ to obtain the map

$$B^2(\mathbb{D}) \otimes_{A(\mathbb{D})} \mathbb{C}_w \xleftarrow{Y \otimes_{A(\mathbb{D})^1} \mathbb{C}_w} \left(pB^2(\mathbb{D})\right) \otimes_{A(\mathbb{D})} \mathbb{C}_w$$

for w in $\mathbb{D}$. In this case we also can identify both Hilbert modules with $\mathbb{C}_w$, but not canonically. Nonetheless, we have the non-negative number $|Y \otimes_{A(\mathbb{D})} 1_{\mathbb{C}_w}|$. Although one can show using exactness that $|Y \otimes_{A(\mathbb{D})} 1_{\mathbb{C}_w}| = 0$ if and only if $p(w) = 0$, it is not true that

$$|Y \otimes_{A(\mathbb{D})} 1_{\mathbb{C}_w}| = |p(w)|$$

unless p(z) is a constant polynomial. In fact, one can show that $|Y \otimes_{A(\mathbb{D})} 1_{\mathbb{C}_w}|$ is not the absolute value of any holomorphic function as was demonstrated in [38]. It is, however, holomorphic — it is a holomorphic section! To see that we must look at the global structure of the localizations, using the sheaf that we introduced in the last chapter.

Both of the Hilbert modules $B^2(\mathbb{D})$ and $pB^2(\mathbb{D})$ define Hermitian holomorphic line bundles over $\mathbb{D}$. Although this was also true for the Hardy module resolution, in this case the two line bundles are not isomorphic as Hermitian holomorphic bundles. It is still the case that the localized map $Y \otimes_{A(\mathbb{D})} 1_{\mathbb{C}_w}$ defines a holomorphic bundle map or, equivalently, a holomorphic section of the bundle Hom $(\underline{pB^2(\mathbb{D})}, \underline{B^2(\mathbb{D})})$. But the "absolute-value" of a holomorphic section does not necessarily have a harmonic logarithm.

Recall that a Hermitian holomorphic bundle E over $\mathbb{D}$ has a natural connection with a curvature two-form $\mathcal{K}_E$ (cf. [38]). For the case of a line bundle the curvature is a complete invariant and can be easily calculated as follows. If $\gamma(w)$ is a non-trivial holomorphic section of E, then

$$\mathcal{K}(w) = \frac{1}{2} \frac{\partial^2}{\partial z \partial \bar{z}} \log \|\gamma(w)\|^2,$$

where we are using $dz d\bar{z}$ as a natural basis for the two-forms on $\mathbb{D}$.

If we let $\gamma(w)$ be a non-vanishing holomorphic section for $\underline{B^2(\mathbb{D})}$, then $\left(Y \otimes_{A(\mathbb{D})} 1_{\mathbb{C}_w}\right)^* \gamma(w)$ is a non-trivial holomorphic section for $\underline{pB^2(\mathbb{D})}$ and a straightforward calculation [56] shows that:

Theorem 6.12. For the polynomial p(z) in $\mathbb{C}[z]$

$$\frac{\partial^2}{\partial w \partial \bar{w}} \log |Y \otimes_{A(\mathbb{D})} {}^1\mathbb{C}_w|^2 = \frac{1}{2}\left\{\mathcal{K}_{B^2(\mathbb{D})}(w) - \mathcal{K}_{pB^2(\mathbb{D})}(w)\right\}.$$

Corollary 6.13. The characteristic operator function $Y \otimes_{A(\mathbb{D})} {}^1\mathbb{C}_w$ is a holomorphic section vanishing precisely at the zeros of p(z) but

$$|Y \otimes_{A(\mathbb{D})} {}^1\mathbb{C}_w| \neq |p(z)|$$

unless p(z) is a constant polynomial.

Although it is possible to calculate the absolute value of $Y \otimes_{A(\mathbb{D})} {}^1\mathbb{C}_w$ directly and show that it is not equal to $|p(z)|$, it is easier in this case to observe that any unitary equivalence between the Hilbert modules $B^2(\mathbb{D})$ and $pB^2(\mathbb{D})$ would extend to a unitary on $L^2(\mathbb{D})$ that commutes with multiplication by z. Therefore, it would have the form M_φ with $|\varphi| = 1$ almost everywhere on all of $\mathbb{D}$, but the assumption that $M_\varphi B^2(\mathbb{D}) = pB^2(\mathbb{D}) \subseteq B^2(\mathbb{D})$ implies that φ is constant from which the result follows.

The preceding results and observations, elementary as they are, make the case that the proper generalization of the notion of "characteristic operator function" is that of a bundle map in a resolution by locally free modules. This makes sense for general resolutions for planar algebras such as those involving the Bergman module for the disk algebra and includes the canonical model theory for the algebras $R(\Omega)$. Moreover, it also makes sense for Hilbert modules for higher dimensional algebras such as $A(\mathbb{D}^n)$ and indicates why it has been difficult to find some valid generalization in an ad hoc manner. But all of these ideas need to be developed before we can be certain that this is the correct approach.

If we take a closer look at the above calculation relating curvatures, then something more is revealed. In calculating the Laplacian of $\log|Y\otimes_{A(\mathbb{D})}1C_w|$ we ignored the zeros of $|Y\otimes_{A(\mathbb{D})}1C_w|$ at which point the logarithm is not defined and hence neither is the Laplacian. Hence, we must interpret the equation in the theorem as being valid only off the zero set. However, there is another way. If we take the distributional Laplacian, then both sides of the equation make sense as distributions but they are no longer equal. One needs to add a distribution to the right hand side equal to a Dirac distribution at each zero counting multiplicity. In other words, there is a correction term telling you information about the quotient module. The same thing occurs in the Hardy module case, where the two curvature terms cancel. What all of this suggests, of course, is an equation for a resolution involving distribution-valued two-forms with a term for each Hilbert module and one for each module map. This becomes much more interesting for Hilbert modules over higher dimensional algebras. We sketch an example

Let $H^2(\mathbb{D}^2)$ be the Hardy module for $A(\mathbb{D}^2)$, let J be the principal ideal in $\mathbb{C}[z_1,z_2]$ generated by $(z_1 - z_2)$ and [J] its closure in $H^2(\mathbb{D}^2)$. As we have indicated earlier, the quotient module $H^2(\mathbb{D}^2)/[J]$ is unitarily equivalent to $d^*B^2(\mathbb{D})$, where d is the diagonal map $d\colon \mathbb{D} \to \mathbb{D}^2$, and $d^*B^2(\mathbb{D})$ is the Bergman module for $A(\mathbb{D})$ pulled back to $A(\mathbb{D}^2)$, or equivalently, the module for $A(\mathbb{D}^2)$ with

$$\varphi\cdot f = (\varphi\cdot d)\cdot f \text{ for } \varphi \text{ in } A(\mathbb{D}^2) \text{ and } f \text{ in } B^2(\mathbb{D}).$$

Thus we have the module resolution

$$0 \leftarrow d^*B^2(\mathbb{D}) \leftarrow H^2(\mathbb{D}) \overset{Z}{\leftarrow} [J] \leftarrow 0.$$

Although it is not obvious, the module [J] is also locally free and hence both sheafs $\underline{H^2(\mathbb{D})}$ and $\underline{[J]}$ are Hermitian holomorphic line bundles over $\mathbb{D}^2$. The sheaf $\underline{d^*B^2(\mathbb{D})}$ is not locally free, however, but is supported on the diagonal $d(\mathbb{D}) \subseteq \mathbb{D}^2$ in $\mathbb{C}^2$. The closure of the diagonal equals $\mathrm{Spec}_{A(\mathbb{D})}(d^*B^2(\mathbb{D}))$. Hence, although $\underline{d^*B^2(\mathbb{D})}$ is not locally free over $\mathbb{D}^2$, it is locally free over the zero variety $d(\mathbb{D})$ of J and thus defines a Hermitian holomorphic vector bundle over it. Further, there is a natural connection on $\underline{d^*B^2(\mathbb{D})}$ with a two-form valued curvature $\mathcal{K}_d$ which can be viewed as a distribution-valued two-form over $\mathbb{D}^2$. One can prove that there exists a smooth function φ, called the regulator, on $d^*\mathbb{D}$ such that

$$\mathcal{K}_{\underline{H^2(\mathbb{D})}} - \mathcal{K}_{[J]} + \varphi\mathcal{K}_\alpha = \tfrac{1}{2}(\mathrm{LAP})\left(\log\|Z\otimes_{A(\mathbb{D}^2)}1_{C_w}\|^2\right)$$

where LAP denotes the distributional, two-form valued Laplacian. While it may be that $\varphi = 1$, it is clear that it is some geometrical invariant.

The context of the preceding discussion could be described as Hermitian algebraic geometry and has been developed in the past for its connections with higher dimensional value distribution theory [27], [49]. More recent work has been carried out in [64] and [65]. The framework studied by the latter authors appears to be very closely related to our program but the source of this connection is not apparent.

In the work of Clark mentioned earlier, his main aim was to calculate invariants for the quotient module defined by the ideal. The zero variety in this case is always one (complex) dimensional and hence the analogue of the preceding discussion applies. The regulator function φ alluded to above can be shown to exist here also and needs to be determined.

As we have indicated earlier, the results described in this chapter are very fragmentary, but what we can see makes it enticing to continue, at least to us. We expect that this is only the beginning.

Bibliography

1. M. B. Abrahamse, *Toeplitz operators in multiply connected regions*, Amer. J. Math. 96 (1974), 261-297.

2. M. B. Abrahamse, *The Pick interpolation theorem for finitely connected domains*, Michigan Math. J. 26(1979), 195-203.

3. M. B. Abrahamse, *Some examples on lifting the commutant of a subnormal operator*, Annales Polonici Math. 37 (1980), 289-298.

4. M. B. Abrahamse and R. G. Douglas, *Operators on multiply connected domains*, Proc. Roy. Irish Acad. Sect. A, 74 (1974), 135-141.

5. M. B. Abrahamse and R. G. Douglas, *A class of subnormal operators related to multiply connected domains*, Advances in Math., 19 (1976), 106-148.

6. M. B. Abrahamse and T. L. Kriete, *The spectral multiplicity of a multiplication operator*, Indiana U. Math. J. 22 (1973), 845-857.

7. J. Agler, *Rational dilation on an annulus*, Ann. of Math. 121 (1985), 537-564.

8. O. P. Agrawal, D. N. Clark and R. G. Douglas, *Invariant subspaces in the polydisk*, Pacific J. Math. 121 (1986), 1-11.

9. O. P. Agrawal and N. Salinas, *Sharp kernels and canonical subspaces*, Amer. J. Math. 109 (1987) 23-40.

10. P. Ahern and D. Clark, *Invariant subspaces and analytic continuation in several variables*, J. Math. Mech. 19 (1969/70), 963-969.

11. T. Ando, *On a pair of commutative contractions*, Acta Sci. Math. 24 (1963), 88-90.

12. W. B. Arveson, *Subalgebras of C^*-algebras*, Acta Math. 123 (1969), 141-224.

13. M. F. Atiyah and I. G. MacDonald, Introduction to Commutative Algebra Addison-Wesley, Menlo Park, California, 1969.

14. S. Axler and P. Bourdon, *Finite codimensional invariant subspaces of Bergman spaces*, Trans. of the Amer. Math. Soc. 305 (1988), 1-13.

15. M. Badri, *On pertubations and products of generalized Bergman kernels*, Thesis, University of Kansas, (1986).

16. M. Badri and N. Salinas, *On products of generalized Bergman kernels*, preprint.

17. M. Badri and P. Szeptycki, *Cauchy products of positive sequences*, preprint (1987).

18. J. A. Ball, *Operators of class C_{00} over multiply connected domains*, Michigan Math. J., 25 (1978), 183-195.

19. J. A. Ball, *A lifting theorem for operator models of finite rank on multiply-connected domains*, J. Operator Theory, 1 (1979), 3-25.

20. J. A. Ball, *Operator extremal problems, expectation operators and applications to operators on multiply connected domains*, J. Operator Theory 1 (1979), 153-175.

21. H. Bercovici, C. Foias, and C. Percy, *Dilation theory and systems of simultaneous equations in the predual of an operator algebra* , Michigan Math J. 30 (1983), 335-354.

22. H. Bercovici, C. Foias, and C. Pearcy, *Dual algebras with applications to invariant subspaces and dilation theory*, CBMS Regional Conf. Ser. in Math., No. 56, Amer. Math. Soc., Providence, (1985).

23. C. A. Berger, L. A. Coburn, and A. Lebow, *Representation and index theory for C^*-algebras generated by commuting isometries*, J. Funct. Anal. 27 (1978), 51-99.

24. C. A. Berger and B. I. Shaw, *Intertwining, analytic structure, and the trace norm estimate*, Proc. Conf. Operator Theory, Springer-Verlag Lecture Notes, Vol. 345 (1973), 1-6.

25. S. Bergman, <u>The Kernel Function and Conformal Mapping</u>, Math Surveys No. 5, Amer. Math. Soc., Providence, R.I., 1950.

26. A. Beurling, *On two problems concerning linear transformations in Hilbert space*, Acta Math. 81 (1949), 239-255.

27. R. Bott and S. S. Chern, *Hermitian vector bundles and the equidistribution of the zeroes of the holomorphic cross-sections*, Acta Math. 114 (1968), 71-112.

28. L. de Branges, *Some Hilbert spaces of analytic functions*, J. Math. Anal. Appl., 11 (1966), 44-72.

29. M. S. Brodskiǐ, <u>Triangular and Jordan representations of linear operators</u>, Nauka, Moscow, 1969; English transl., Transl. Math. Monographs, Vol. 32, Amer. Math. Soc. Providence, R.I., (1971).

30. L. G. Brown, R. G. Douglas, and P. A. Fillmore, *Unitary equivalence modulo the compact operators and extensions of C^*-algebras*, Proc. Conf. Operator Theory, Springer-Verlag Lecture Notes, Vol.345 (1973), 58-128.

31. R. W. Carey and J. D. Pincus, *Principal functions, index theory, geometric measure theory and function algebras*, Integral Equations and Operator Theory, 2 (1979), 441-483.

32. J. L. Casti, <u>Nonlinear system theory</u>, Mathematics in Science and Engineering, vol. 175, Academic Press, New York, (1985).

33. C. T. Chen, <u>Introduction to linear systems theory</u>, Holt, Rinehart, and Winston, NewYork, (1970).

34. D. N. Clark, *On commuting contractions*, J. Math. Anal. Appl., 32 (1970), 590-596.

35. D. N. Clark, *Commutants that do not dilate*, Proc. Amer. Math. Soc., 35 (1972), 483-486.

36. D. N. Clark, *Restrictions of H^p functions in the polydisk*, preprint.

37. A. Connes, *Noncommutative differential geometry*, Publ. IHES 86(1986), 41-144.

38. M. J. Cowen and R. G. Douglas, *Complex geometry and operator theory*, Acta Math. 141 (1978), 187-261.

39. M. J. Cowen and R. G. Douglas, *On operators possessing an open set of eigenvalues*, Memorial Conf for Féjer-Riesz, Budapest, 1980, Colloq. Math. Soc. J. Bolyai, (1980), 323-341.

40. M. J. Cowen and R. G. Douglas, *On moduli for invariant subspaces*, Operator Theory: Advances and Applications, vol. 6, pp. 65-73, Birkhauser Verlag, Basel, (1982).

41. M. J. Crabb and A. M. Davie, *von Neumann's inequality for Hilbert space operators*, Bull. London Math. Soc. 7 (1975), 49-50.

42. R. Curto and D. Herrero, *On closures of joint similarity orbits*, Integral Equations and Operator Theory, 8 (1985), 489-556.

43. R. Curto and P. Muhly, *C^*-algebras of multiplication operators on Bergman spaces*, J. Funct. Anal. 64 (1985), 315-329.

44. R. Curto, P. Muhly, and J. Xia, *Hyponormal pairs of commuting operators*, Operator Theory: Advances dual Applications, Vol. 38, Birkhauser Verlag, Basel (1988), 1-22.

45. R. Curto and N. Salinas, *Generalized Bergman kernels and the Cowen-Douglas theory*, Amer. J. Math. 106 (1984), 447-488.

46. J. Dixmier, *C^*-algebras*, North-Holland, New York, (1977).

47. P. G. Dixon, *The von Neumann inequality for polynomials of degree greater than two*, J. London Math. Soc. (2), 14 (1976), 369-375.

48. P. G. Dixon and S. W. Drury, *Unitary dilations, polynomial identities and the von Neumann inequality*, Math. Proc. Camb. Phil. Soc. 99 (1986), 115-122.

49. S. Donaldson, *Anti-self-dual Yang-Mills connections over complex algebraic surfaces and stable vector bundles*, Proc. London Math. Soc.(3), 50 (1985), 1-26.

50. R. G. Douglas, *Banach Algebra Techniques in Operator Theory*, Academic Press, New York, (1972).

51. R. G. Douglas, *Hilbert modules for function algebras*, Operator Theory: Advances and Applications, vol. 17, Birkhauser Verlag, Basel, (1986), pp. 125-139.

52. R. G. Douglas, *Hilbert modules for function algebras*, Szechuan Lectures, preprint, (1986).

53. R. G. Douglas, *On Šilov resolutions of Hilbert modules*, in Operator Theory: Advances and Applications, V. 28, Birkhauser Verlag, Basel, (1988) 51-60.

54. R. G. Douglas and C. Foias, *A homological view in dilation theory*, preprint (1976).

55. R. G. Douglas and C. Foias, *Subisometric dilations and the commutant lifting theorem*, *Operator Theory: Advances and Applications*, Vol. 12, Birkhauser Berlag, Basel, (1984), 129-139.

56. R. G. Douglas and G. Misra, *Some calculations for Hilbert modules*, preprint (1986).

57. R. G. Douglas and V. I. Paulsen, *Completely bounded maps and hypo-Dirichlet algebras*, Acta Sci. Math., 50 (1986), 143-157.

58. R. G. Douglas and K. Yan, *Rigidity of Hardy submodules*, preprint (1988).

59. S. W. Drury, *Remarks on von Neumann's inequality*, Proceedings of a Special Year in Harmonic Analysis, Lecture Notes in Mathematics, Vol. 995, Springer-Verlag, Berlin, 14-32.

60. N. Dunford and J. T. Schwartz, *Linear operators I: General theory*, Interscience, New York, (1958).

61. P. Fillmore, *Notes on operator theory*, van Nostrand, New York, (1970).

62. S. D. Fisher, *Function theory on planar domains*, Wiley - Interscience publications, John Wiley & Sons, New York, (1983).

63. T. W. Gamelin, *Uniform Algebras*, Prentice Hall, Engelwood Cliffs, N. J., (1969).

64. H. Gillet and C. Soule, *Intersection sur les varietes d'Arakelov*, C. R. Acad. Sci. Paris, v. 299, ser. I, 12 (1984), 563-566.

65. H. Gillet and C. Soule, *Classes caracteristiques en theorie d'Arakelov, C. R. Acad. Sci. Paris*, v. 301, ser. I, 9 (1985), 439-442.

66. I. Glicksberg, *Function algebras with closed restrictions*, Proc. Amer. Math. Soc. 14 (1963), 158-161.

68. P. R. Halmos, *Shifts on Hilbert spaces*, J. Reine Angew. Math. 208 (1961), 102-112.

69. P. R. Halmos, *Ten problems in Hilbert space*, Bull. Amer. Math. Soc. 76 (1970), 887-993.

71. H. Helson, *Lectures on invariant subspaces*, Academic Press, New York (1964).

72. H. Helson and D. Lowdenslager, *Prediction theory and Fourier series in several variables*, Acta Math. 99 (1958), 165-202.

73. J. W. Helton and R. Howe, *Traces of commutators of integral operators*, Acta Math. 136 (1976), 271-305.

74. D. A. Herrero, *A Rota universal model for operators with multiply connected spectrum*, Rev. Roum. Math. Pures et Appl. 21 (1976), 15-23.

75. D. A. Herrero, *Approximation of Hilbert space operators*, Pitman, Boston, (1982).

76. R. V. Kadison, *On the orthogonalization of operator representations*, Amer. J. Math. 77 (1955), 600-620.

77. P. D. Lax, *Translation invariant spaces*, Acta Math. 101 (1959), 163-178.

78. M. S. Livšic, *Operators, oscillations, waves. Open systems*, Nauka, Moscow, 1966; English transl., Transl. Math. Monographs, vol. 34, Amer. Math. Soc., Providence, R. I., (1973).

79. G. Misra, *Curvature inequalities and extremal properties of bundle shifts*, J. Operator Theory 11 (1984), 305-318.

80. G. Misra and N. S. N. Sastry, *Completely bounded modules and associated extremal problems*, preprint (1987).

81. W. Mlak, *Decompositions and extensions of operator valued representations of function algebras*, Acta Sci. Math. (Szeged) 30 (1969), 181-193.

82. W. Mlak, *Decompositions of operator valued representations of function algebras*, Studia Math. 36 (1960), 111-123.

83. W. Mlak, *Commutants of subnormal operators*, Bull. Acad. Polon. Sci. 19 (1971), 837-842.

84. W. Mlak, *Intertwining operators*, Studia Math. 43 (1972), 219-233.

85. W. Mlak, *Operator valued representations of function algebras*, ISNM vol. 25 Birkhauser Verlag, Basel and Stuttgart, (1974), 49-79

86. J. von Neumann, *Allgemeine Eigenwert-theorie Hermitesches Funktional Operatoren*, Math. Ann. 102 (1929), 49-131.

87. J. von Neumann, *Eine spektraltheorie für allgemeine Operatoren eines unitären Raumes*, Math. Nachr. 4 (1951), 258-281.

88. S. K. Parrott, *Unitary dilations for commuting contractions*, Pacific J. Math. 34 (1970), 481-490.

89. V. I. Paulsen, *Every completely polynomially bounded operator is similar to a contraction*, J. Funct. Anal. 55 (1984), 1-17.

90. V. I. Paulsen, Completely Bounded Maps and Dilations, Pitman Research Notes in Mathematics, vol. 146, Longman, London, (1986).

91. V. I. Paulsen, *K-spectral values for some finite matrices*, J. Operator Theory, 18 (1987), 249-264.

92. V. I. Paulsen, *Subnormal models and completely positive maps*, preprint, (1986).

93. G. C. Rota, *On models for linear operators*, Comm. Pure Appl. Math. 13 (1960), 468-472.

94. W. Rudin, Function Theory in Polydisks, Benjamin, New York, (1969).

95. N. Salinas, *Products of kernel functions and module tensor products*, preprint, Integral Equations and Operator Theory 11 (1988).

96. D. Sarason, The H^P spaces of an annulus, Memoirs Amer. Math. Soc., vol. 56, Providence, R. I., (1965).

97. D. Sarason, *On spectral sets having connected complement*, Acta Sci. Math. 26 (1965), 289-299.

98. D. Sarason, *Generalized interpolation in* H^{∞}, Trans. Amer. Math. Soc. 127 (1969), 179-203.

99. E. L. Stout, The Theory of Uniform Algebras, Bogden and Quigley, Tarrytown-on-Hudson, New York, (1971).

100. B. Sz.-Nagy, *Sur les contractions de l'espace de Hilbert*, Acta Sci. Math. 15 (1953), 87-92.

101. B. Sz.-Nagy and C. Foias, Harmonic analysis of operators on Hilbert Space, American Elsevier, New York, (1970).

102. J. Taylor, *A joint spectrum for several commuting operators*, J. Funct. Anal. 6 (1970), 172-191.

103. J. Taylor, *A general framework for a multi-operator functional calculus*, Advances in Math. 9 (1972), 183-252.

104. N. Th. Varopoulos, *On an inequality of von Neumann and an application of the metric theory of tensor products to operators theory*, J. Funct. Anal. 16 (1974), 83-100.

105. D. Voiculescu, *Norm-limits of algebraic operators*, Rev. Roum. Math. Pures et Appl. 19 (1974), 371-378.

106. H. Wold, A study in the analysis of stationary time series, Stockholm, (1938).

107. K. Yan, *Equivalence of some principal submodules*, preprint, (1988).

108. R. G. Douglas, V. Paulsen, and K. Yan, *Operator Theory and Algebraic Geometry*, Bull. Amer. Math. Soc. 20 (1989), 1-5.

Index